introduction to ENVIRONMENTAL SCIENCE *with Lab*

KIMBERLY L. FRYE
MARGARET A. WORKMAN
DePaul University

Kendall Hunt
publishing company

Interior images: Photos on page 14 © 2012 by Kimberly L. Frye
Photo on page 25 © 2012 by Kimberly L. Frye

Cover images:
Soil Quality Testing © 2012 by Margaret A. Workman
Water Quality Testing and Rooftop Garden © 2012 by Kimberly L. Frye
Urban Atmosphere © 2012 by Shutterstock, Inc.

www.kendallhunt.com
Send all inquiries to:
4050 Westmark Drive
Dubuque, IA 52004-1840

CONTENTS

LABORATORY 1

Scientific Measurements

OBJECTIVES

1. Be able to measure length and height using the metric system
2. Be able to relate two variables and identify patterns
3. Be able to create scatter graphs, and evaluate relationship strengths between two variables

INTRODUCTION

Science is an attempt to discover order in nature (i.e., cause and effect); unfortunately, our universe is not a very orderly place! Trying to make sense of all the random variables and discover a method within the madness is difficult. The way most scientists wrestle with the problem of identifying patterns of cause and effect in an orderly fashion is through using the *scientific method*.

- The first thing scientists must do is ask a question or identify a problem to be investigated.
- Then scientists working on this problem collect *scientific data* by making *observations* and taking measurements.
- The primary goal of science is not just the data but a new idea, principle, or model that connects and explains the data, which can used **to make useful predictions about what should happen in nature**. Scientists working on a particular problem try to come up with a variety of possible explanations, or *scientific hypotheses*, of what they observe in nature. To be accepted, a scientific hypothesis not only must explain scientific data and phenomena but also should make predictions that can be used to test the validity of the hypothesis.
- Once a scientific hypothesis is determined, experiments are conducted to test the predictions. Experiments can support or not support various hypotheses, but they can **never prove** that any hypothesis is the best or the only explanation.

In this laboratory, you will:
1. Become familiar with the metric system

2. Make observations
3. Construct hypotheses
4. Make measurements
5. Test hypotheses
6. Evaluate data

EXERCISE 1: HYPOTHESIS TESTING

You have probably observed that tall people have longer arms, whereas shorter people have shorter arms. This is an observation about patterns that exist between the bones that compose our skeleton. Certified as a science by the FBI in the 1930s, forensic anthropology has determined that size and shape of a bone can determine the age, sex, height, and build of an individual; for example, the length of an arm or leg bone found as remains can be used to calculate the height of an individual. Such calculations are based on equations created from large amounts of data collected about bone characteristics that are then graphed as data points to reveal patterns about the relationship between particular bones. Today we will work with the specific relationship between the upper arm bone, humerus, and height.

Hypothesis: The length of a humerus bone (in cm) is related to body height (in cm).
- You will measure your partner's humerus and height, and then record your data on the board.
- **You must perform each measurement; do not just ask for measurements!**
- Record class data on the data sheet on Table 1.1.

1. According to your table, can you see a relationship between the humerus and height measurements (i.e., can you make an educated guess as to whether the height is 4x the humerus? 10x?)

2. Now we will construct an x-y scatter plot of humerus length (cm) and height (cm) using Microsoft Excel in order to gather information about the relationship between our data points *(each data point = one person's humerus AND height)*. <u>Include a copy of the graph with your report.</u>

   - Which measurement belongs on the x-axis (humerus or height)?

   - Why was that measurement assigned to the x-axis?

   - What is the equation of the line?

   - Which measurement does **x** represent in your equation?

   - Which measurement does **y** represent in your equation?

   - What is the y-intercept? (Be sure to define and give the number from the equation.)

   - What is the correlation of determination (R^2)? What does this mean? (Be sure to define and give the number from the equation.)

Laboratory 1: Scientific Measurements

3. Now let's use the equation generated by our data to see how accurately it can predict height from a humerus measurement:
  - Take your humerus measurement and plug it into the equation our class generated (*show your work below*).

  - Does your calculated height match your actual height measurement?

  - How close are the two numbers?

  - Now let's separate the data to make two graphs for male and female. What are the equations for the two lines?

  - Now try plugging in your humerus measurement into the equation generated by the appropriate line (male or female, so if you are male use the male line equation…).

  - Is this calculated value any closer to your actual height measurement compared with values before we separated male and female data?

EXERCISE 2: CONSTRUCTING A HYPOTHESIS

In this exercise you are free to come up with any hypothesis you wish to test. The two variables you investigate must be numeric and must be features that you can measure. Check your hypothesis with the instructor or T.A. before proceeding to data collection.

1. What observation are you basing your hypothesis on?

2. What is your hypothesis?

3. Describe how you will collect data. (Use Table 1.2 to collect at least 20 data points.)

4. Make an x-y scatter plot of your data. What is the equation of the line?

5. What is the correlation of determination (R^2)?

6. Did the data support or fail to support your hypothesis? Explain.

7. What are possible sources of error in your testing procedure?

8. How would you revise this hypothesis and testing procedure?

TABLE 1.1. *Testing the Relationship between the Humerus and Height*

STUDENT	HEIGHT (CM)	LENGTH OF HUMERUS (CM)	MALE OR FEMALE
1			
2			
3			
4			
5			
6			
7			
8			
9			
10			
11			
12			
13			
14			
15			
16			
17			
18			
19			
20			
21			
22			
23			
24			
25			
26			
27			
28			
29			
30			
31			
32			
33			
34			
35			

Laboratory 1: Scientific Measurements

TABLE 1.2. *Constructing a Hypothesis*

1		
2		
3		
4		
5		
6		
7		
8		
9		
10		
11		
12		
13		
14		
15		
16		
17		
18		
19		
20		
21		
22		
23		
24		
25		
26		
27		
28		
29		
30		
31		
32		
33		
34		
35		

Laboratory 1: *Scientific Measurements*

LABORATORY 2

Human Population Survivorship Curves

OBJECTIVES

1. Understand survivorship curves
2. Understand differences in human mortality between pre-1900 and post-1900 cohorts

INTRODUCTION

Life expectancy for humans has increased significantly over the past 100 years. With industrialization came better nutrition, medical care, and technology that helped increase human life expectancy. In 1900, the U.S. life expectancy was around 49 years. In 2011, it is 78 years. This increase in life expectancy has caused an increase in the population size. Particular noteworthy is the decline in child mortality over the past 100 years.

Survivorship curves are often used to study population trends. A survivorship curve is a graph that shows the proportion of individuals surviving at each age for a given cohort. We generally classify survivorship curves into three categories: Type I, Type II and Type III (See Figure. 2.1). In Type I, most individuals in a population are lost when they are older. In Type II, there is an equal amount of loss at all ages. In Type III, most individuals die when young. In this lab, you will examine trends in human population growth rates and survivorship in the Chicago Region.

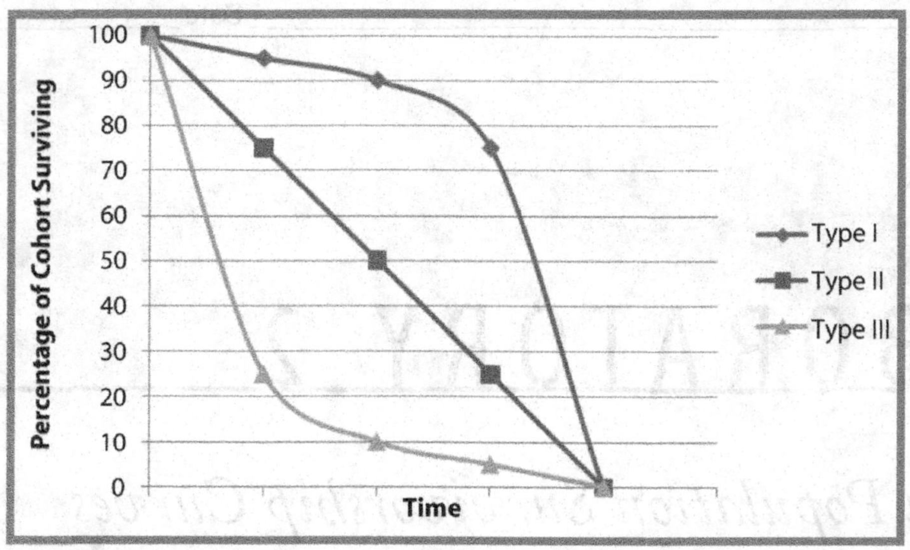

FIGURE 2.1. *Idealized Survivorship Curves*

PROCEDURE

1. Collect Data

You will be collecting data from two sources: a cemetery and newspaper obituaries. We will take a field trip to a local cemetery and record the age of death for males and females who lived their life completely <u>before</u> 1910. This will be your "historic" cohort. (If inclement weather, use data from http://www.findagrave.com/cgi-bin/fg.cgi?page=cs) Then we will return to the classroom and record the age of death for males and females who lived their life completely <u>after</u> 1910. This will be your "modern" cohort. (If no hard copies of obituaries, use online obituaries at http://www.legacy.com/obituaries/chicagotribune/#&did=Obits&lpos=Sub). <u>You must collect 20 males and 20 females for EACH cohort.</u>

Hints:
- If you cannot determine from the name whether it is a male or female, do not use that data point.
- If you are working with a partner, do not collect the same data points.

On the attached DATA SHEETS, make a "tally mark" in the appropriate cell (e.g. ⋕⋕). Please pay close attention that you are in the correct column (Cemetery vs. Obituary) and that you are in the right column (Male vs. Female).

2. Create a Class Data Set

After completing your individual data collection, we will put everyone's data together to make one big, class data set. This will result in better survivorship curves. Your instructor will have an Excel spreadsheet set up in the classroom. When you are finished with your data collection, enter your data TOTALS for EACH cohort (historic males, historic females, modern males, modern females) for EACH age group (i.e., total up your tally marks in each cell and enter that total in the appropriate cell of the master classroom data set). The Excel sheet will look something like this:

Laboratory 2: Human Population Survivorship Curves

AGE OF DEATH	STUDENT 1	STUDENT 2	STUDENT 3	STUDENT 4	CLASS TOTAL
0					
0-1					
1-5					
5-10					

NOTE THERE WILL BE A SEPARATE SPREADSHEET FOR EACH OF THE 4 COHORTS! Enter your data on the appropriate sheet. Your instructor will assemble the data and give you the numbers for the class totals. RECORD THIS CLASS TOTAL ON YOUR RESULTS TABLES IN THE "Number Dying" COLUMN. (See page 11.)

3. Create the Survivorship Curves

To determine a survivorship curve use the following method: Since you will be collecting data on historic males, historic females, modern males, and modern females, you can determine the number surviving to each age for the four groups by using the following technique. <u>Use the class data to fill in the results table for all four cohorts.</u>

<u>Example – Assume a total sample size of 600.</u>

TABLE 2.2. *Cemetery: All Four Cohorts*

AGE OF DEATH (YEARS)	NUMBER DYING	NUMBER SURVIVING	PERCENT SURVIVING
0	0	600	(600/600) * 100% = 100%
0-1	30	600 – 30= 570	(570/600) * 100% = 95%
1-45	50	570 – 50 = 520	(520/600) * 100% = 87%
5-10	80	520 – 80 = 440	(440/600) * 100% = 73%

Hints:
- The number surviving each cohort should go down with each age category. You should end up with "0" in the last age category.
- When calculating the percent surviving, you always divide by the total sample size. You should end up with "0" in the last age category.

<u>Use the class data to graph a survivorship curve for each of the four groups:</u> historic females; historic males; modern females; modern males. Place all four of these graphs onto one sheet to print out and turn in with your report. To make the graph, set up an Excel sheet that looks like Table 2.3.

AGE AT DEATH (YEARS)	HISTORIC MALES (% SURVIVING)	HISTORIC FEMALES (% SURVIVING)	MODERN MALES (% SURVIVING)	MODERN FEMALES (% SURVIVING)
0				
0-1				
1-5				
5-10				

On the y-axis should be the % surviving and on the x-axis should be the age categories. Remember to use good graph presentation techniques by labeling axes, labeling legend, and so on.

Student Data Sheet

AGE AT DEATH (YEARS)	CEMETERY		OBITUARY	
	MALE	FEMALE	MALE	FEMALE
0-1				
1-5				
5-10				
10-15				
15-20				
20-25				
25-30				
30-35				
35-40				
40-45				
45-50				
50-55				
55-60				
60-65				
65-70				
70-75				
75-80				
80-85				
85-90				
90-95				
95-100				
100+				

Results Table: Class Results

AGE AT DEATH (YEARS)	CEMETERY						OBITUARY					
	MALE			FEMALE			MALE			FEMALE		
	NUMBER DYING	NUMBER SURVIVING	PERCENT SURVIVING	NUMBER DYING	NUMBER SURVIVING	PERCENT SURVIVING	NUMBER DYING	NUMBER SURVIVING	PERCENT SURVIVING	NUMBER DYING	NUMBER SURVIVING	PERCENT SURVIVING
0-1												
1-5												
5-10												
10-15												
15-20												
20-25												
25-30												
30-35												
35-40												
40-45												
45-50												
50-55												
55-60												
60-65												
65-70												
70-75												
75-80												
80-85												
85-90												
90-95												
95-100												
100+												

1. Describe the similarities and differences you see among the four graphs.

2. Compare the child mortality between the historic and modern cohorts. What factors might account for any difference?

3. Compare the mortality for adults in their reproductive ages (20-40). Do you see differences between the sexes? Do you see differences between the historic and modern cohorts? What factors might account for any differences?

4. Compare the mortality for adults between 60 and 80. Do you see differences between the sexes? Do you see differences between the historic and modern cohorts? What factors might account for any differences?

5. Compare the mortality for adults over 80. Do you see differences between the sexes? Do you see any differences between the historic and modern cohorts? What factors might account for any differences?

6. Why might data such as this be useful?

LABORATORY 3

Natural Selection

OBJECTIVES

1. Understand how environmental changes alter phenotypic frequency in populations
2. Be able to create a line graph with multiple data series

INTRODUCTION

This lab demonstrates the process of natural selection. We will see how organisms cope with changes in the environment by noticing how the environment directs the preferential survival of certain traits in the populations over others. Specifically we perform a simulation of a population of fictional birds (that are the same species) as they try to survive through generations after changes in their environment occur that alter food sources, food availability time, landscape, and vegetation.

BACKGROUND

Pretend that you are a member of a population of seed-eating birds. A great deal of genetic variation exists in this population. We'll focus on the variation seen in the beak shapes of these birds. Because of the different shapes of their beaks (phenotypes), some birds in this population eat in different ways. <u>These birds are all the same species; they are a population with genetic variation!</u> This is similar to the way humans are all the same species but have genetic variations that cause different eye color.

There are many different types of beaks observed in this population of birds. Phenotypes that are observed in this population of birds include tweezers, spoons, clothespins, skewers, pencils, and straws (Figure 3.1).

FIGURE 3.1. *Different Phenotypes Present in Our Population of Birds*

These birds get all of their nourishment from eating certain food. Right now, conditions are good and there are many types of food available in this environment (cereal, cat food, marshmallows, raisins).

Though these birds can eat any of these foods to survive, some birds prefer to eat certain things. This is because birds with particular beak shapes have an easier time obtaining and eating certain foods, and under certain environmental conditions. Think about the types of food we will use in this simulation and the types of beaks in the population: can you make an educated guess about which beak type you would expect to perform the best?

PROCEDURE

1. The instructor will assign certain students to be team captains. Team captains will be in charge of keeping track of the number of food pieces obtained by their team members.

2. The instructor will assign additional students to work with each team captain. All teams must have the same number of students. Extra students will have the chance to participate during later hunts.

3. Each group of students represents a group of birds having the same beak phenotype (feeding mechanism). Each member of the group should get:
   • One of the feeding mechanisms (beak) [same for entire group]
   • A "stomach" (baggie)

4. You and your group will try to get food using your "beak" during five rounds of play. You will be competing with the other groups of birds to capture food. The instructor will throw the food into the feeding area. You should face away from the hunting area while the food is being thrown out. When the instructor gives the signal, you will try to capture food as quickly as you can until STOP is called.

   Rules of the Hunt
   - Only one beak per bird.
   - Food must be lifted with the feeding mechanism (beak) and placed into the "stomach" (baggie) held in the opposite hand.
   - You must lift the food into the stomach. You can't drag the baggie along the ground to capture food.
   - You are permitted to cooperate with other members of your beak type.
   - Cheating is possible! You may take food from another bird's beak; however, food in the baggie that has been "eaten" cannot be stolen.

5. After the STOP signal is given:
   a. Count how many pieces of food you have collected.
   b. Tell your team captain how many pieces you have collected.
   c. Return your food to the food collection container set out by the instructor.
   d. Wait while your captain calculates how many pieces your team has gathered all together and reports this to the instructor.

6. If you are a captain:
   a. Add up how many pieces of food each team member has captured. Record the number on Table 3.1 (both on your worksheet (pg. 17) and on the board).
   b. The instructor will add up the totals reported by each team, to determine a class total. Each captain will use that total to calculate the number of players earned for his/her team for the next generation by using the formula below:

$$\text{Total players for your team in the next round} = \frac{\text{Total pieces captured by your team}}{\text{Total pieces captured by the class}} \times \text{total \# of students playing}$$

   c. Report your number of players for the next round on Table 3.1 (both on your worksheet and on the board).
   d. If your team gathered relatively little food and thus earned fewer players than you started with, some players must turn in their beaks and join the group of extras. If your team collected a lot of food and thus earned more players than you started with, additional players from the group of extras may join you.
   e. To begin the next round of play (the next generation), give new members of your team feeding mechanisms (beaks). This symbolizes the birth of babies having the same trait as their parents.

RESULTS

1. Complete Table 3.1, which shows the amount of food obtained and how many players each team earned during each of the five generations (rounds of play). Students must record this information in their lab write-up. First, write in the number of players each group began with in the column labeled "Generation 0." Next, write in the number of pieces of food each group gathered in each round and the number of members of each generation.

2. Fill in Table 3.2 (pg. 18). First, copy the number of birds in each group from Table 3.1. Next, calculate the percentages of each beak shape in the total population over the five generations. To find the percent of the population having a particular beak shape for each generation, first:
   a. Add up the total number of players for each round.
   b. Divide the total number of survivors from each group by the total number of people playing.

3. Graph the results, showing how the percentages of beak shapes in the total population changed over time. Each line will represent data for a beak type. All groups will be shown on the same graph, so you will use the "series" function on excel. Be sure to label both the x and y-axes. Use the x-axis for time (generation) and the y-axis for percentage of the total population.

GROUP	GENERATION 0	GENERATION 1 30 SEC. FEEDING TIME NO VEGETATION		GENERATION 2 45 SEC. FEEDING TIME		GENERATION 3 30 SEC. FEEDING TIME FAMINE: CEREAL ONLY		GENERATION 4 30 SEC. FEEDING TIME VEGETATION SPROUTS!		GENERATION 5 30 SEC. FEEDING TIME EARTHQUAKE!	
	# OF PLAYERS	# FOOD PIECES	# OF SURVIVORS	# FOOD PIECES	# OF SURVIVORS	# FOOD PIECES	# OF SURVIVORS	# FOOD PIECES	# OF SURVIVORS	# FOOD PIECES	# SURVIVORS
spoon											
tweezers											
clothespin											
skewer											
pencil											
straw											
Total											

TABLE 3.2. *Results*

GROUP	GENERATION 0		GENERATION 1		GENERATION 2		GENERATION 3		GENERATION 4		GENERATION 5	
	# OF SURVIVORS	% TOTAL POP.	# OF SURVIVORS	% TOTAL POP.	# OF SURVIVORS	% TOTAL POP.	# OF SURVIVORS	% TOTAL POP.	# OF SURVIVORS	% TOTAL POP.	# OF SURVIVORS	% TOTAL POP.
spoon												
tweezers												
clothespin												
skewer												
pencil												
straw												
Total												

QUESTIONS

1. Suppose that the cat food is made in the shape of an X and became the only food source. Write a hypothesis as to which of the beak phenotypes would be "naturally selected."

2. Imagine that the changes in the environment had made it so that only cereal remained for the remainder of the game. Predict how this might change the percentages of bird beak phenotypes seen in the population after five generations. Would these results differ from the results we got in this demonstration? Why or why not?

3. Remember that each group is a sub-set of the entire population of imaginary birds. These sub-groups represent variations in beak phenotypes in a single species. It is important to note that even if your group died out during the simulation, the population as a whole survived by hunting cat food. However, the overall genetic variation in the population was reduced. How does genetic diversity (having a lot of genetic variation) allow a species to better survive in a changing environment?

LABORATORY 4

Using GIS to Investigate Urban Forestry

OBJECTIVES

1. To understand the role of urban forestry in urban land management
2. To learn how to generate data using GPS (global positioning system)
3. To learn how to make a file using a GIS (geographical information system)

INTRODUCTION

Urban forests broadly include urban parks, street trees, landscaped boulevards, public gardens, river and coastal promenades, greenways, river corridors, wetlands, nature preserves, natural areas, shelter belts of trees, and working trees at industrial Brownfield sites.[1] Urban trees have a history that begins with trees as landscape embellishment. Today urban trees have come to be seen as essential components of city infrastructure, as critical to human life as food, housing, and other public utilities. Urban trees are now valued for the ecosystem services that they provide (e.g. preventing erosion, air pollutant removal, oxygen, shade, etc.). Yet to efficiently make use of these benefits, trees must reach maturity, as leaf number and size directly affects the tree's ability to provide ecosystem services. Urban forestry has had to develop its own forestry methods to address the needs and challenges unique to urban trees as compared to their woodland counterparts.

Can you think of some challenges that urban trees face that woodland trees do not?

The following excerpt from the USDA Forest Service illustrates the urban tree perspective and policies of federal government.

1. United States Department of Agriculture (USDA) Forest Service. Accessed October 2, 2009. http://www.fs.fed.us/ucf/program.html. Last modified October 22, 2008 by http://www.fs.fed.us/.
2. Ibid.

Benefits of urban forests:
Urban forests are dynamic ecosystems that provide needed environmental services by cleaning air and water, helping to control stormwater, and conserving energy. They add form, structure, beauty, and breathing room to urban design, reduce noise, separate incompatible uses, provide places to recreate, strengthen social cohesion, leverage community revitalization, and add economic value to our communities.

Interconnected systems:
Urban forests, through planned connections of green spaces, form the green infrastructure system on which communities depend. Green infrastructure (http://www.greeninfrastructure.net/) works at multiple scales, from the neighborhood to the metro area up to the regional landscape. This natural life support system sustains clean air and water, biodiversity, habitat, and nesting and travel corridors for wildlife, and connects people to nature.[2]

The management of urban trees is an interdisciplinary practice involving architecture, landscaping, planning, development, horticulture, and so on. One particular discipline involved in forestry is geography, especially through the use of geographical information systems (GIS). GIS allows for extensive data collection and management through ever-improving user interfaces (increasing the user-friendly quality of very comprehensive sets of information that can be used by many individuals across many agencies!). Today you will practice forestry techniques of identifying tree species while collecting geographical data to record your tree locations with GPS (global positioning system) using a handheld GPS receiver. You will also use Google Earth to record your GPS work in a format that can be shared publicly.

PROCEDURE

GPS via Handheld GPS Receiver

1. Work in groups of 3.

2. You will be assigned a group number. Once you have your number, look at the map provided to your group (and on the large map drawn on the board) to find the location assigned to your group number.

3. Write your group names on the board by your assigned group number.

4. Be sure your group has all materials listed below before leaving the room.
   Materials:
   - 1 clipboard
   - 2 datasheets
   - 1 GPS handheld number – *Turn on to ensure the batteries are working.*
        Write down the number written on your GPS handheld unit.
   - Field guide and notes

5. We will all proceed outside of the building for a quick tutorial on using your GPS unit to get longitude/latitude information through the Mark or Waypoint tool.

6. Then your group should proceed to your assigned location and begin to identify each tree species in your area.

7. At each tree use your data sheets to note the:
    - **Species** – use your field guide and notes
    - **Postal location** – any visible or discernable postal address (e.g. 1253 W. Lill)
    - **Longitude/Latitude location** using your GPS unit – although the unit records this information, _**you want to also write it down before marking the next tree;**_ this will allow for easier access to the information during the lab wrap-up when we're indoors and out of range of all satellite communication.

If you are unsure about a particular species, ask your instructor or T.A. before labeling it inaccurately. For any tree that remains "unknown," make as many notes as you can, take a photo if you can, and grab a leaf/bark/flower/fruit/twig sample.

When you have completed gathering all data for your area, return to the lab and any computer that has Google Earth (e.g. your laptop or any provided computer on campus). Your group will then create a map of your trees using the instructions below for Google Earth.

GIS VIA GOOGLE EARTH INSTRUCTIONS

1. Type in your tree coordinates, save them in the "My Places" folder using the "add placemark" feature, and name each tree by its common name (e.g. Norway Maple, Honey Locust, etc.).

2. **Using the "Add Placemark" tool:** Once they are all placemarked and named, right-click on the My Places name, select "Save As," and save the folder with all your placemarks into the My Documents folder on the computer.

3. **Saving the My Places Folder (right-click menu):** Name your file using your GROUP number to identify your group. Then someone in the group needs to attach the kmz file to an e-mail and send to your instructor or teaching assistant.

4. All kmz files will be compiled to create a tree map of our study area. Our map will be added to the existing forestry database maintained by DePaul's Urban Forestry Laboratory.

DATA COLLECTION SHEET

GROUP #: **GROUP NAMES:** **DATE:**

KEY

H.L. = HONEY LOCUST
ASH = ASH
C.A. = CRABAPPLE
S.M. = SILVER MAPLE
BIRCH = BIRCH
N.M. = NORWAY MAPLE
HAWTHORN = HAWTHORN
LINDEN = LINDEN
REDWOOD = REDWOOD
ELM = ELM

TREE NO.	SPECIES	BUILDING NO.	STREET	LONGITUDE/LATITUDE & EXTRA NOTES

Name ____ Date ____________________

LABORATORY 5

Statistical Analysis of Earthworm Populations

OBJECTIVES

1. Understand the role of invasion in ecosystems
2. Understand how to use a bar graph to compare sample populations
3. Understand how to use statistics to compare sample populations

INTRODUCTION

Invasive species are an increasing threat to the biodiversity of an ecosystem. Earthworm invasion has only recently been investigated as a major factor in so-called "invasional meltdowns." An invasional meltdown is the process where one invasion of a species facilitates the invasion of others (Belote and Jones 2009). Thus, the rate of loss of ecological health can greatly accelerate as one invasive species makes way for additional ones.

A study conducted by Lee E. Frelich et al (2006) claims earthworms "can be considered a keystone class of organisms that exert control over many aspects of ecosystem structure and function, including the primary producers" (Frelich et al 2006, p.1236). Earthworms are invading much of the continent of North America. Recent research provides evidence that they have been colonizing the cold-temperate region of North America since the days of European settlement. Their invasion of the woodlands has caused declines in diversity and abundance of native understory plants (Hale et al 2005, Frelich et al 2006). The native earthworms of this region were most likely annihilated by the Wisconsin glaciation approximately 12,000 years ago (Madritch and Lindroth 2009). This open niche was filled when the European earthworms were brought over with settlement and trade. Earthworms have been negatively impacting the ecosystems that they have invaded. As so-called "ecosystem engineers," they have a great ability to modify habitats and change community species composition (Belote and Jones 2009). They have been credited with the reduction of the thickness of the organic layers exposing mineral soil, increasing of soil bulk densities, increasing nutrient cycling rates leading to an overall decrease in essential nutrient availability such as N (nitrogen) and P (phosphorous), impacting seedling survival, and the disrupting plant hyphal networks (Frelich et al 2006, Belote and Jones 2009, Hale et al 2005, Madritch and Lindroth 2009).

The region of the Midwest is of particular importance in the study of invasive earthworms. Much of this region previously was without earthworms, and the impact of the invasive earthworms has been pronounced. It has been reported that the invasion of earthworms in northern forests of Minnesota has caused significant declines in the diversity and cover of herbaceous plants and abundance of seedlings (Frelich et al 2006). After invasion, the organic matter in the forest floor is reduced and species are lost. In addition to this, there is a relationship between invasive earthworms and at least some invasive shrubs. Heneghan et al (2007) proposed that the impact of earthworms on litter breakdown creates conditions that promote and sustain invasion by the invasive shrub *Rhamnus cathartica* (European Buckthorn).

There has been evidence that above-ground disturbance impacts earthworm populations below-ground: "Intermediate frequency or intensity of disturbance will maximize diversity. At high disturbance frequency, species diversity is predicted to be low, because only 'weedy' species that quickly colonize and reach maturity are able to survive" (Hughes, 2010). Comparing a highly manipulated landscape (e.g., well managed lawn) to a less manipulated site (e.g., a park that receives little management or a vacant lot) offers a chance to determine if human activity (disturbance) is correlated with European earthworm populations, as invasive species research suggests.

After collecting our data, we will use a bar graph (aka a column chart) to compare worm abundance between sites. We will also use a statistical measure known as the "student's t-test" to determine if there is a statistically significant difference in worm abundance between the two collection sites. The student's t-test was published in 1908 by William Sealy Gosset, a chemist in the Guinness, to assess population means by using small samples. He chose to publish under the pseudonym "Student" because Guinness considered the statistical test to be a trade secret (Raju, 2005)!

PROCEDURE

Your group will collect three samples from each location (undisturbed and disturbed) using the mustard extraction method. At your first sample site:

1. Pour all of your mustard solution into your water bottle.

2. Pour some liquid back into the empty mustard bottle, shake, and re-pour contents back into the water bottle.

3. Shake your water bottle vigorously to mix.

4. Remove all loose leaves and twigs from your sample spot.

 Laboratory 5: Statistical Analysis of Earthworm Populations

5. Pour onto your sample spot approximately 1/3 of the volume from your water bottle.

6. Start your timer and wait 2-3 minutes or until the mustard solution is completely absorbed into the ground.

7. Collect any European earthworms that come up.

8. Pour out just a little more solution, wait for solution to be absorbed, and collect for 1-2 minutes more.

Repeat steps 5-8 for each of your three sample sites. Repeat all steps at your second location for another three sample sites. Take your collections back to the lab to complete the data worksheet on the next page.

REFERENCES

Belote, R.T., R.H. Jones. 2009. Tree leaf litter composition and nonnative earthworms influence plant invasion in experimental forest floor mesocosms. *Biological Invasions* 11:1045-1052

Frelich, L.E., C.M. Hale, S. Scheu, A.R. Holdsworth, L. Heneghan, P.J. Bohlen, P.B Reich. 2006. Earthworm invasion into previously earthworm-free temperate and boreal forests. *Biological Invasions* 8:1235-1245

Hale, C.M., L.E. Frelich, P.B. Reich. 2006. Changes in hardwood forest understory plant communities in response to European earthworm invasions. *Ecology*, 87:1637-1649

Hale, C.M, L.E Frelich, P.B. Reich, J. Pastor. 2005. Effects of European earthworm invasion on soil characteristics in Northern Hardwood Forests of Minnesota, USA. *Ecosystems* 8:911-927

Heneghan, L., J. Steffen, K. Fagen. 2007. Interactions of an introduced shrub and introduced earthworms in an Illinois urban woodland: an Impact on leaf litter decomposition. *Pedobiologia* 50:543-551

Hughes, A. (2010) Disturbance and Diversity: An Ecological Chicken and Egg Problem. *Nature Education Knowledge* 1(8):26

Madritch, M.D., R.L Lindroth. 2009. Removal of invasive shrubs reduces exotic earthworm populations. *Biological Invasions* 11:663-671

Pavao-Zuckerman, Mitchell A. 2008. The Nature of Urban Soils and Their Role in Ecological Restoration in Cities. *Restoration Ecology* 16.4: 642-649

Raju TN (2005). "William Sealy Gosset and William A. Silverman: two "students" of science." *Pediatrics* 116 (3): 732-5

Hypothesis:

Manipulated (high management/disturbed) site—collect 3 samples

How many total worms did you collect in your first location? ____________

How many total worms did you collect in your second location? ____________

How many total worms did you collect in your third location? ____________

Total number of worms for this site: ____________

Total number of worms divided by 3 (mean of worm abundance): ____________

Report your mean number of worms for the manipulated site on the class data table.

Un-manipulated (little to no management/undisturbed) site—collect 3 samples

How many total worms did you collect in your first location? ____________

How many total worms did you collect in your second location? ____________

How many total worms did you collect in your third location? ____________

Total *number* of worms for this site: ____________

Total *number* of worms divided by 3 (mean of worm abundance): ____________

Report your mean number of worms for the un-manipulated site on the class data table.

ANALYSIS

1. Once the data for the entire class is compiled, calculate the class mean number of worms and standard deviations for each collection site.

2. Use the class means to create a bar graph of worm abundance at each site. Use your calculated standard deviations to generate error bars on each bar.

3. Use the class data to run a student's t-test to determine if there is a statistically significant difference in worm abundance between the two collection sites.

1. What kind of graph are we using to see if there is a difference in worms collected between the manipulated and un-manipulated sites?

2. What measure are we using to determine how much variation we have within our collections?

3. How do we represent variation within collections on this type of graph?

4. What are we using statistics to do in this lab?

5. What is the statistical test we are using to analyze our worm data?

6. What is the alpha value? (Give *both* the actual value and a brief definition.)

7. What measure do we compare to the alpha? (Give *both* the name of the number and the actual value.)

8. How do we use the above comparison (between our generated statistical measure and the alpha value) to analyze our hypothesis?

9. Is there a statistical difference between the manipulated and un-manipulated sites?

   Laboratory 5: *Statistical Analysis of Earthworm Populations*

LABORATORY 6

Soil Testing

OBJECTIVES

1. Demonstrate ability to perform soil quality assessment techniques
2. Demonstrate ability to use a bar graph to compare sample populations
3. Analyze statistics to compare sample populations

INTRODUCTION

According to the USDA, soil can be defined as "the unconsolidated mineral or organic material on the immediate surface of the Earth that serves as a natural medium for the growth of land plants." While most people do not think twice about soil, except maybe when it gets on their clothes, it provides an invaluable service. Soil helps control water flow, sustains plants, filters pollutants, and cycles nutrients. All growing plants need soil. Soil helps plants grow in two ways: it provides a place for plant roots to get a good, strong grip, and it stores nutrients and water that the plants use to make food.

There are many different kinds of soil. Soils form through the interactions of climate, living organisms, and landscape position as they influence the decomposition and transformation of parent material over time.

PHYSICAL CHARACTERISTICS OF SOIL

Soil Texture

Over a long period of time, rocks get broken into smaller and smaller pieces by weathering, erosion, water, and animals, and turn into soil. The size of these soil particles affects how easily both water and air can pass through the soil to reach the roots of plants. Soil particles can be separated into three separate categories: sand, silt, and clay. **Soil texture** describes the proportions of sand, silt, and clay in a soil.

- **Sand**–These particles are biggest (0.05-2 mm). Because they are so big, there are lots of air spaces between the sand particles, so water passes quickly through the soil before the plant can absorb it.

- **Silt**–These particles are smaller than sand particles (0.002-0.05 mm). Because they are smaller, there is less room between them and water can't move as quickly through the soil. The water is trapped for a longer time next to the plant roots, and they can absorb it.

- **Clay**–These particles are the smallest (< 0.002 mm). They can be packed close together, so the spaces for water to pass through the soil are very small. Soils with some clay particles are able to store water and nutrients for a long time, which is important for growing plants. But soils with too much clay are very dense and heavy. This makes it difficult for water and plant roots to push through the soil.

Soil mixtures that have similar distributions of sand, silt, and clay, will have similar properties and are, therefore, grouped into the same soil class. Twelve soil textural classes are recognized and their compositions are designated on a textural triangle (see Figure 6.1). Soil class names consist of the terms sand, silt, clay and loam used as nouns, adjectives, or both. When the percentages of sand, silt, and clay in a soil sample are known (i.e. when tests are done in the lab), the class name can be determined by plotting theses values on a textural triangle. Soil texture affects the suitability of soil for building and construction use, urban planning, and pollution control.

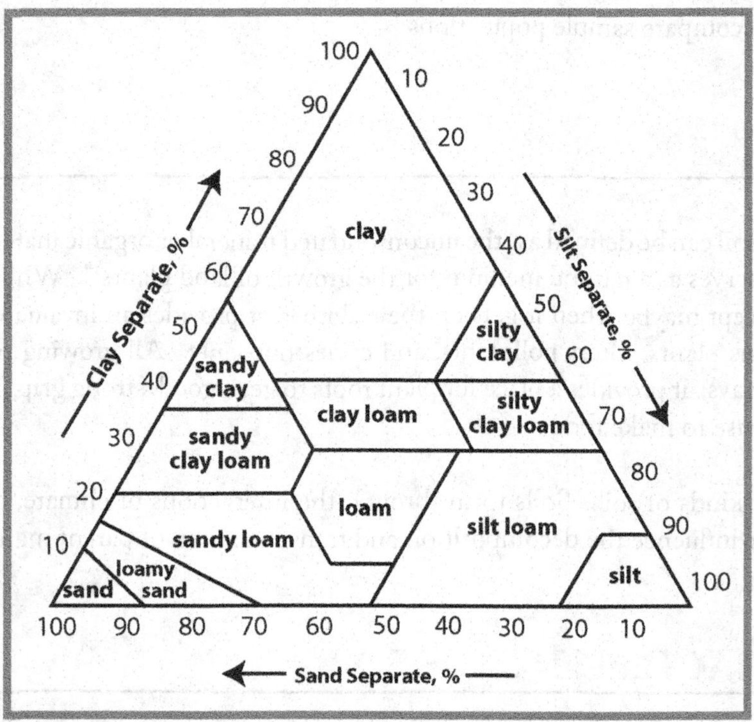

FIGURE 6.1. *Textural Triangle (Source: United States Department of Agriculture)*

Soil Color

Soil color can indicate a soil's stage of development or mineral origin. Soil colors are most conveniently measured by comparison with a color chart. The collection of charts generally used with soils is a modified version of a Book of Color and includes only that portion needed for soils. The color charts in the book display color chips systematically arranged. The arrangement is by the three dimensions of Hue, Value, and Chroma that help distinguish color. The Hue notation of a color indicates its relation to Red, Yellow, Green, Blue, and Purple; the Value notion indicates its lightness; and the Chroma notation indicates its strength.

The colors displayed on each page of the Soil Color Charts are of constant Hue, designated by a symbol in the upper right hand corner of the card.

Vertically the colors become successively lighter from the bottom of the card to the top; their value increases. Horizontally they increase in Chroma from left to right. The Value notation of each chip is indicated by the vertical scale in the far left column of the chart. The Chroma notation is indicated by the horizontal scale across the bottom of the chart.

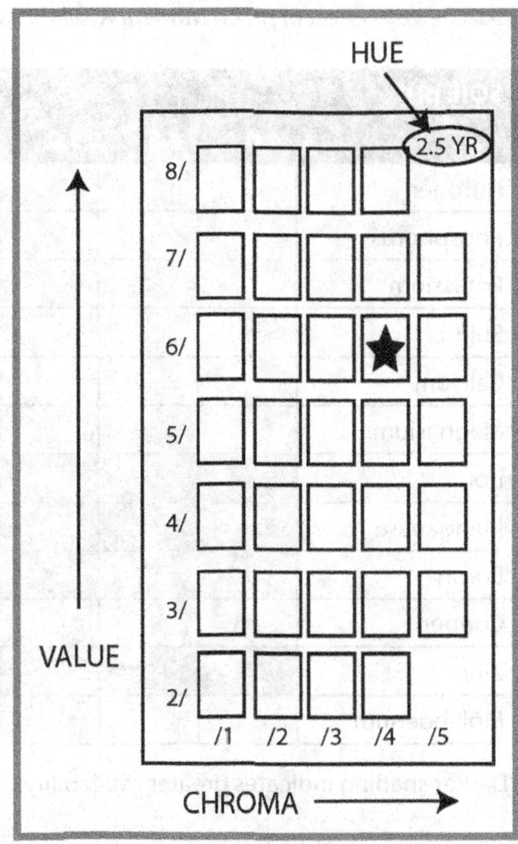

FIGURE 6.2. *Soil Color Chart*

When writing the scientifically correct name for soil color, the order is "hue, value, chroma" with a space between the hue letter and the succeeding value number, and a diagonal between the two numbers for value and chroma. Thus the notation for a color of hue 2.5YR, value 6, and chroma 4, is 2.5YR 6/4 (see Figure 6.2).

CHEMICAL PROPERTIES OF SOIL

pH

pH is a measure of how acidic or alkaline a soil is. The pH of soil is measured on a scale from 0 to 14 (See Figure 6.3). Zero is the most acidic kind of soil and 14 is the most alkaline. Soil that is just right for most plants is equally acidic and alkaline. It

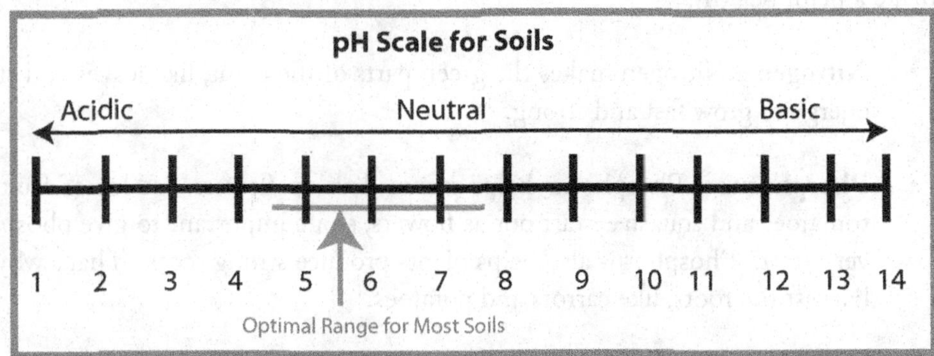

FIGURE 6.3. *pH Scale for Soils*

has a pH around 7 on the pH scale. Some plants, like strawberries, like soil that is a little bit acidic. Others, like tomatoes, like a soil that is slightly alkaline.

FIGURE 6.4. *Effect of pH on Nutrient Availability*

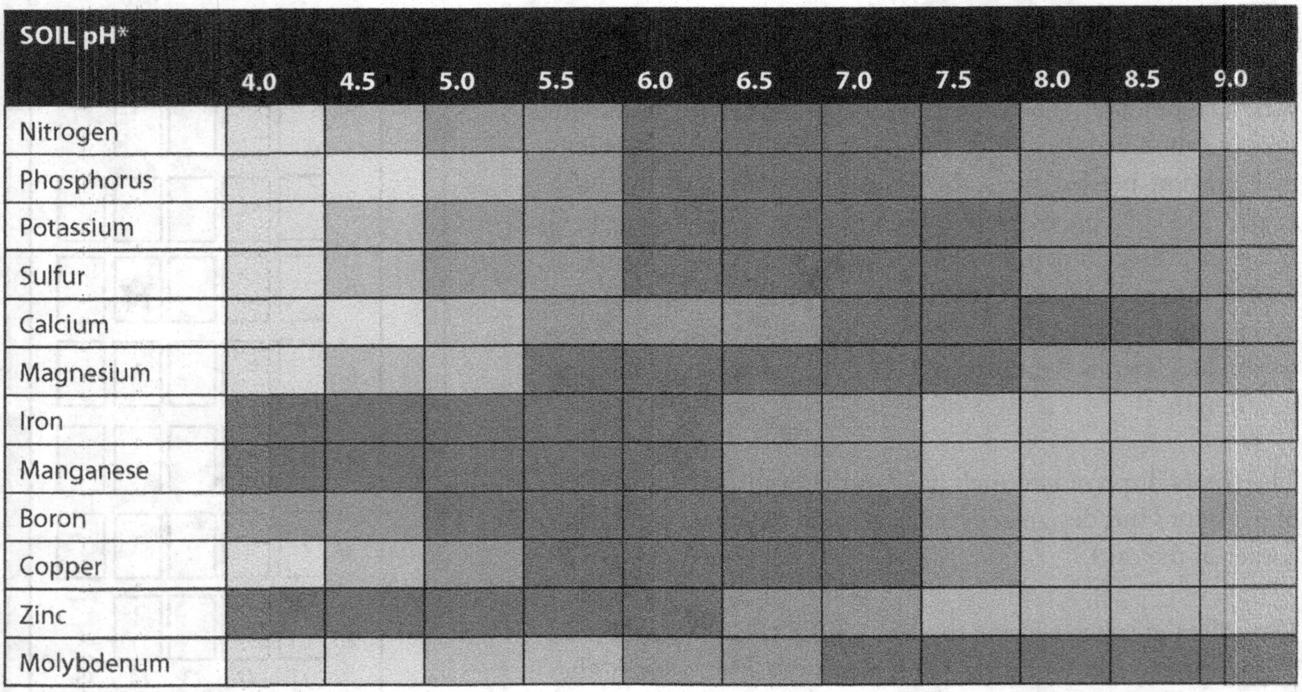
SOIL pH*
4.0
4.5
5.0
5.5
6.0
6.5
7.0
7.5
8.0
8.5
9.0
Nitrogen
Phosphorus
Potassium
Sulfur
Calcium
Magnesium
Iron
Manganese
Boron
Copper
Zinc
Molybdenum

*Darker shading indicates greater availability

Soil pH plays a big role in the availability of nutrients to plants. For example, potassium is more available at a higher soil pH range (more alkaline) than when the pH is lower. Conversely, iron is more available at the lower soil pH range (more acidic).

Nutrients

Plants need air, light, and water to grow…and they need nutrients. Nutrients are like "vitamins" for healthy plants. Plants absorb nutrients that are dissolved in water through their roots. The three most important nutrients found in soil are nitrogen (N), phosphorus (P) and potassium (K). Each nutrient has a special job to make a plant healthy.

- **Nitrogen** – Nitrogen makes the green parts of the plant, like leaves and stems, grow. It also gives plants energy to grow fast and strong.

- **Phosphorus** – Phosphorus helps plants develop flowers. A lot of fruits and vegetables—like apples, tomatoes and squash—start out as flowers, so its important to give phosphorus to those plants from the very start. Phosphorus also helps plants produce strong roots. That's why it is critical for plants that we harvest the roots, like carrots and potatoes.

- **Potassium** – Potassium is needed for new growth, like root tips and buds, and it helps the plant to fight diseases.

Most plant food contains the nutrients nitrogen, phosphorus and potassium. Some plants have special needs, so there are plant foods like rose food and African Violet food. The numbers on the container tell the percentage of nitrogen, phosphorus, and potassium that are in the plant food. For example, 15–30–15 plant food has 15% nitrogen, 30% phosphorus, and 15% potassium.

PROCEDURES

In this lab activity, we will be investigating the differences and similarities between urban soil taken from the parkway, and garden soil that has been enriched to increase production of vegetables. You will be assigned either the urban soil or the garden soil and will perform two physical and four chemical tests on the soil.

> Physical Tests: Soil Texture and Soil Color
> Chemical Tests: pH, Nitrogen, Phosphorus and Potassium

Each member of your group should perform at least one trial for each of the tests. Record your group's data on the Data Table (pg. 38) and also on the board, since we will be calculating a class average for each test for each soil type.

Soil Texture

It is possible to identify the texture of a soil by simply handling a moist sample. This method is used whenever exact proportions are not needed or when in the field. It requires considerable practice to become an expert; however, the novice can quickly obtain satisfactory results by practicing the following the rubric.

1. Follow the instructions for "Determining Soil Texture by the Feel Method" found at the Soil Texture Lab Station.

2. Record your result on the data table.

3. Other members of the group should repeat the procedure.

Soil Color

1. Find the book of soil colors at the Soil Color Lab Station.

2. Find a hunk of soil the size of a marble.

3. Place the soil behind the openings on the pages of the color book until you find the one it most closely resembles.

4. Write down the name of the soil color, as described in the soil color section of the introductory material, on the data table (including Hue, Value, and Chroma).

5. Each member of the group should independently perform this procedure.

pH

1. Follow directions for Aqueous Extraction Method as found at the Soil pH Lab Station.

2. Follow directions for Measuring pH of above aqueous extract as found at the "pH Station." You do not need to filter the extract.

3. Record pH on your data table.

4. Each member of the group should determine the pH of your soil.

Nitrogen

1. Follow directions for Calcium Sulfate Extraction for Soil as found at the Soil Nitrogen Lab Station.

2. Follow directions for Measuring Nitrate-Nitrogen in soil as found at the Soil Nitrogen Lab Station.

3. Record amount of nitrate-nitrogen in the soil on your data table.

4. Repeat for each member of the group.

Phosphorus

1. Follow directions for Mehlich 2 Extraction for Soil as found at the Soil Phosphorus Lab Station.

2. Follow directions for Measuring Phosphorus in Soil as found at the Soil Phosphorus Lab Station.

3. Record amount of phosphorus in the soil on your data table.

4. Each member of the group should perform this test.

Potassium

1. You may use the <u>extract from the Mehlich 2 Extraction for Soil</u> that you produced above in the Phosphorus procedure (Step 1) for this test.

2. Follow directions for Measuring Potassium in Soil as found at the Soil Potassium Lab Station.

3. Record amount of potassium in the soil on your data table.

4. Each member of the group should perform this test.

ANALYSIS

1. The instructor will tell you which type of soil you have (urban vs. garden) based on your sample unknown number.

2. Write your measured values on the board under the correct heading.

3. Once the data for the entire class is compiled, calculate the class average and standard deviation for each chemical test.

4. Use the class averages to create a bar graph of the values for pH, nitrogen, phosphorus, and potassium. Use your calculated standard deviations to generate error bars on each bar. <u>Include a copy of your graph with your report.</u>

5. Use the class data to run a student's t-test to determine if there is a statistically significant difference in concentrations between the two types of soil (urban vs. garden) for each chemical test (pH, nitrogen, phosphorus and potassium).

DATA TABLE

Soil Unknown Sample # _____Urban_____

Hypothesis: ____

Soil Texture:
    Trial 1 ____
    Trial 2 ____
    Trial 3 ____

Soil Color:
    Trial 1 ____
    Trial 2 ____
    Trial 3 ____

pH:
    Trial 1 ____
    Trial 2 ____
    Trial 3 ____

    AVE _________________________________

Nitrate-Nitrogen:
    Trial 1 ____
    Trial 2 ____
    Trial 3 ____

    AVE ___________ x 2 = _______________

Phosphorus:
    Trial 1 ____
    Trial 2 ____
    Trial 3 ____

    AVE ___________ x 3.3 = _____________

Potassium:
    Trial 1 ____
    Trial 2 ____
    Trial 3 ____

    AVE ____________________

Class Results

	TEXTURE	COLOR	pH	NITRATE	PHOSPHORUS	POTASSIUM
Urban Soil						
Garden Soil						
Urban Ave. + SD						
Garden Ave. + SD						

1. Based on your statistical analysis, is there a difference between the urban and the garden soil for any of the chemical tests?

2. Compare and contrast the urban soil with the garden soil. Make sure you discuss all of the parameters you tested. Also make sure to include a comparison of the averages and the SD.

3. What might account for the differences that are observed?

LABORATORY 7

Genetically Modified Foods

OBJECTIVES

1. Demonstrate ability to perform DNA extraction and purification
2. Perform a polymerase chain reaction (PCR)
3. Analyze a gel electrophoresis
4. Assess whether a food product was genetically modified

INTRODUCTION

The "Green Revolution" of the 1970s allowed us to feed our growing population through the use of high-yield varieties of grain, synthetic fertilizers, and pesticides. However, with continuing population growth and negative environmental impacts of fertilizers and pesticides, we are looking for other ways to increase food production. One such way is through genetic modification of crop plants. This allows us to increase nutrition and decrease our use of pesticides. However, these techniques are relatively new and (some would argue) the impacts are not well understood.

The process involves directly manipulating an organism's genetic material in the laboratory by adding or deleting genes. Genes that encode herbicide and/or insect resistance and also drought/frost tolerance are routinely added to the DNA of commercial plants. In 2010, 86% of the U.S. corn harvest and 93% of soybean and cotton crops were genetically modified!

One notable gene that is commonly inserted into crops is the glycophosate-resistance gene that protects food crops against the herbicide Roundup. Plants that have had this gene inserted into their DNA are resistant to the herbicide Roundup. So this herbicide can be sprayed on the crops in order to kill weeds without any impact on the crop itself. This genetic modification is advantageous because it allows for better weed control, higher yields, and lower environmental impact.

In this lab, we will isolate DNA from a food product (corn or soybean) and use polymerase chain reaction (PCR) to look for evidence of the 35S promoter that drives the expression of the glycophosate-resistance gene. If the 35S promoter is found, the food product was made from plants that were genetically modified. Amplification of tubulin is also necessary for the success of this lab. Tubulin is found in all plants, so it will serve as a control in the experiment. If evidence of tubulin is found, you know that you performed the experiment correctly and actually isolated some DNA and processed it correctly.

We will be using a scientific process called polymerase chain reaction (PCR) in this laboratory activity. This process takes a single piece of DNA and makes thousands (to millions) of copies of it. This is done with thermal cycling, which is repeated cycles of heating to "unzip" the DNA and cooling to "rezip" the DNA with new base pairs to form new copies of the original DNA. The amplified (copied) DNA needs to be made in sufficient numbers in order for it to be seen on an agarose gel. The agarose gel will separate the DNA according to size (number of base pairs). The glycophosate-resistant gene that we are interested in will appear around 162 bp (base pairs).

PROCEDURE

Part I: Isolating DNA from Your Food Product

1. You will be assigned a non-GM control or a GM control sample. Cut 2 pieces of the soybean leaf about ¼ inch long. Put it in a 1.5mL tube and label with control type (non-GM or GM) plus your name.

2. Take your dry food that contains soy or corn and place it in a plastic bag. Seal the bag. Crush the dry food with a mallet until it is a powder. Add enough to a 1.5mL tube until it reaches half-way up to the .1mL mark. Label with your name.

3. Add 100μL of Buffer A to EACH tube.

4. Grind for 1 minute with a plastic pestle. Use a new, clean one for each tube so you do not contaminate either sample.

5. Add 900μL of Buffer A to each tube.

6. Vortex the tubes for 5 seconds.

7. Boil the sample for 5 minutes.

8. Centrifuge the tubes for 2 minutes.

9. Transfer 350μL of each of the LIQUIDS to new tubes. Discard the old tubes with the solid material left behind.

10. Add 400μL of isopropanol to each of the new tubes.

Laboratory 7: Genetically Modified Foods

11. Invert tubes a few times and then let sit for 3 minutes.

12. Place tubes in centrifuge with caps facing out. Centrifuge for 5 minutes.

13. Pour off liquid in the tubes and discard THE LIQUID. Remove remaining liquid with a 100µL pipet.

14. Let the tubes sit with caps off for 10 minutes to allow the remaining isopropanol to evaporate.

15. Add 100µL of Buffer B to each tube and dissolve the solid pellet.

16. Let sit at room temperature for 5 minutes.

17. Centrifuge tubes for 1 minute.

18. Keep the tubes on ice while you continue working.

Part II: Amplify DNA by PCR

1. Get 2 PCR tubes containing PCR beads. Label with your name.

2. Label one tube as "35S Food" and the other as "35S non-GM" or "35S GM," depending on which control group you were assigned. These will be used for the 35S promoter reactions.

3. Add 22.5µL of 35S dye to each tube.

4. Put a clean tip on your pipet and then put 2.5µL of your food solution from Part I into the tube labeled as "35S Food."

5. Put a clean tip on your pipet and then put 2.5µL of your "non-GM" or "GM" control group solution from Part I into the tube labeled as "35S non-GM" or "35S GM."

6. Get 2 more PCR tubes containing PCR beads. Label with your name.

7. Label one tube as "T Food" and the other as "T non-GM" or "T GM," depending on which control group you were assigned. These will be used for the tubulin reactions.

8. Add 22.5µL of tubulin dye to each tube.

9. Put a clean tip on your pipet and then put 2.5µL of your food solution from Part I into the tube labeled as "T Food."

10. Put a clean tip on your pipet and then put 2.5µL of your "non-GM" or "GM" control group solution from Part I into the tube labeled as "T non-GM" or "T GM."

11. Keep your samples on ice until you are ready to begin the thermal cycling.

12. Program the thermal cycler for 32 cycles of the following:

Denaturing Step: 94°C for 30 seconds
Annealing Step: 60°C for 30 seconds
Copying Step: 72°C for 30 seconds

Have the program end with a 4°C hold pattern after the 32 cycles are finished.

13. Store the DNA at -20°C until ready for next step.

Part III: Gel Electrophoresis

1. Gels will be prepared for you ahead of time.

2. Put a clean tip on your pipet and then put 20µL of each of your samples into different wells on the 2% agarose gel according to the following layout:

Marker	Non-GM		GM	
pBR322/ BstNI	Tubulin	35S	Tubulin	35S

Marker	Food Group 1		Food Group 2	
pBR322/ BstNI	Tubulin	35S	Tubulin	35S

3. Put 20µL of the pBR322/BstNI Marker into one well on the gel.

4. Run the gels at 130V for 30 minutes.

Part IV: Staining the Gels

1. Remove the gel and place it in a staining tray.

2. Cover the gel with the staining solution and allow it to sit for 15-20 minutes. Shake the tray GENTLY.

3. Pour the stain back into the bottle. It can be reused (8 times).

4. Cover the gel with DI water (do not use tap water). Shake the tray GENTLY.

5. Change the DI water every 10 minutes until all the stain is removed (up to 40 minutes).

6. View the gel on a lightbox.

7. Photograph the gel. **Include a copy of the photograph of your gel with your report.**

Part V: Interpretation of Gels

1. Orient the photograph of the gel containing your sample so that the wells are at the top.

2. Look at your gel to notice patterns. You should see one or two prominent bands in each lane.

3. Locate the lane that contains the Marker pBR322/BstNI. It will be on the left-hand side of the gel. Working from the top, locate each of the following bands:

1857 bp (base pair)	large size
1058 bp	
929 bp	
383 bp	
121 bp	small size

    This lane will serve as a kind of scale for your gel.

4. The Tubulin gene, if present, will appear around 187 bp. The 35S promoter, if present, will appear around 162 bp. Use the Marker pBR322/BstNI as your scale. The tubulin and 35S, if present, should both appear between the 121 bp and 383 bp of the Marker.

5. If the tubulin bands appear on your gel, that means you did the experiment correctly and some DNA was amplified. If it does not appear, something went wrong.

6. If the 35S promoter band appears on your gel, that means that the DNA was genetically modified. If it does not appear, the DNA was not genetically modified.

1. What is the purpose of the tubulin wells on the gel?

2. What is the purpose of the Marker (pBR322/BstNI) well on the gel?

3. What was the purpose of performing the PCR reaction on the non-GM soybean leaf?

4. What was the purpose of performing the PCR reaction on the GM soybean leaf?

5. Analyze the following FICTITIOUS gels, and identify whether Food 1 and/or Food 2 were genetically modified.

Food 1 _______________________   Food 2 _______________________

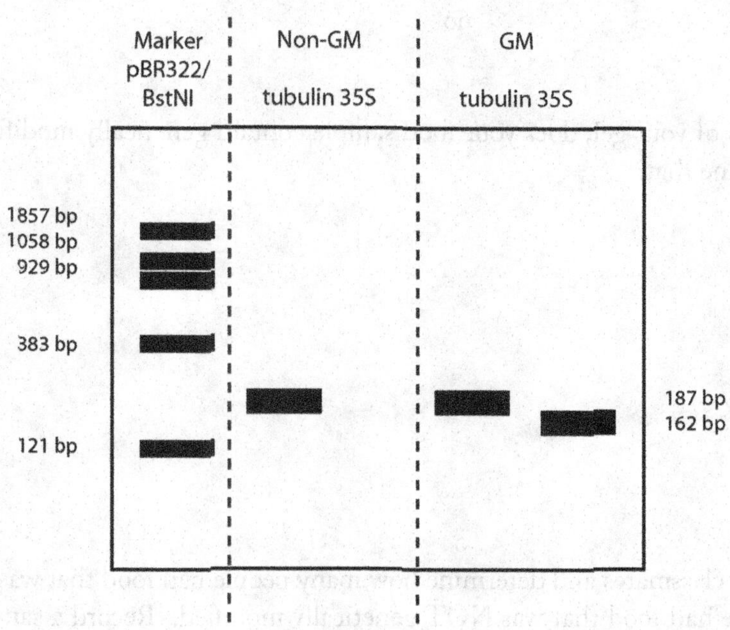

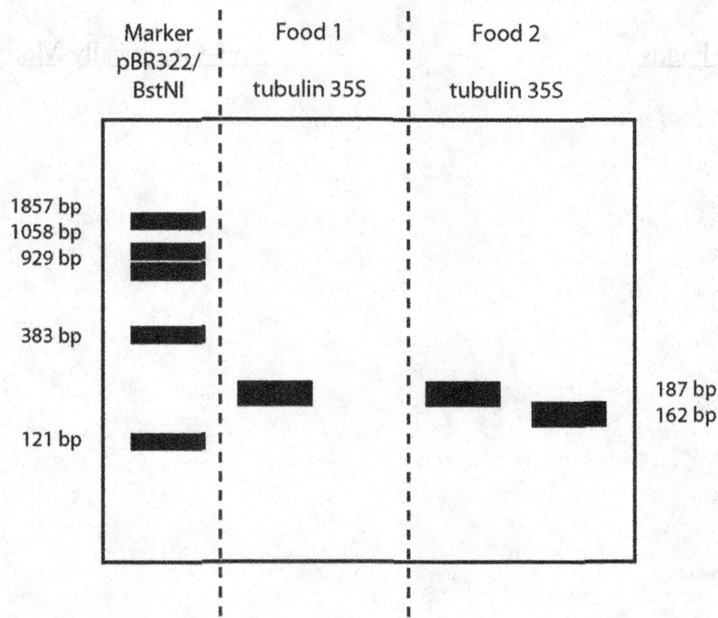

6. Look at your gel and determine whether the following bands appear for your food product:

Tubulin: yes   no

35S: yes   no

7. Based on the analysis of your gel, does your food sample contain genetically modified soybeans or corn? How did you determine that?

8. Take a survey of your classmates and determine how many people had food that was genetically modified and how many people had food that was NOT genetically modified. Record a sampling of products for each category. Be sure to include Brand Name and Item Name.

<u>Genetically Modified Foods</u>   <u>Non-Genetically Modified Foods</u>

                                   Laboratory 7: *Genetically Modified Foods*

LABORATORY 8

Biofuels: Producing Ethanol from Various Feedstocks

OBJECTIVES

1. Understand how yeast metabolizes sugar to produce ethanol
2. Understand how to use a gas sensor to monitor CO_2 concentrations
3. Determine the rate of yeast respiration with various feedstocks (sugar, starch and cellulosic material)
4. Determine the impact of enzymes on the fermentation of cellulosic material

INTRODUCTION

The United States is looking to wean itself off of fossil fuels, especially petroleum used in gasoline. Global climate change, dependence on foreign countries, and increased political instability around the world are only a few reasons why. One possible way to decrease dependence on petroleum as a transportation fuel is by using more ethanol. Currently, regular gasoline contains between 5.9 and 10% ethanol as an additive. Special flex-fuel vehicles can use E85 gasoline, which is 85% ethanol. In fact, in the Energy Independence and Security Act of 2007, the U.S. government mandated that the total amount of biofuels added to gasoline would increase from just 4.7 billion gallons in 2007 to 36 billion gallons by 2022.

In Brazil, ethanol is made by using sugarcane as the feedstock. The main product of sugarcane is sucrose, which is a disaccharide of glucose and fructose. The glucose ($C_6H_{12}O_6$) is separated, then fermented by yeast to produce ethanol and carbon dioxide. The chemical reaction is as follows:

$$C_6H_{12}O_6 \longrightarrow 2\,CH_3CH_2OH + 2\,CO_2$$
$$\text{yeast}$$

However, in the U.S., we make ethanol from corn. This requires more effort than making ethanol from sugarcane. To make ethanol from corn, we first need to grind up the corn kernels. We then treat the ground corn with enzymes to convert the starch in the corn to glucose. After this step, the process continues as in the sugarcane method above by using yeast to ferment the glucose into ethanol and carbon dioxide. The chemical reaction is as follows:

$$(C_6H_{12}O_6)_{300\text{-}600} \xrightarrow{\text{enzyme}} C_6H_{12}O_6 \xrightarrow{\text{yeast}} 2\ CH_3CH_2OH + 2\ CO_2$$

The production of ethanol from corn is problematic, however. For one thing, it takes some corn out of the food supply, thus driving up prices. It is also energy- and fertilizer-intensive to produce corn, decreasing its desirability as a transportation fuel alternative to petroleum. So, scientists are turning to cellulosic material to make ethanol. These materials are wood, grasses, and non-edible parts of plants. These are more desirable because they do not take food out of the food supply. The problem is that, in order to release the glucose from the cellulosic material, much more effort is needed, because the glucose from cellulosic material is bound up in cellulose, which is then wrapped with hemicelluloses and lignin. First, the cellulose needs to be extracted from the hemicelluloses and lignin bindings. This is done through a pretreatment of grinding and acid hydrolysis. Then, the cellulose is treated with enzymes to break it up into its component glucose. Finally, the glucose can be fermented with yeast to produce ethanol and water. The process can be summarized as follows:

$$\text{Cellulosic material} \xrightarrow{\text{pretreatment}} (C_6H_{12}O_5)_n \xrightarrow{\text{enzyme}} C_6H_{12}O_6 \xrightarrow{\text{yeast}} 2\ CH_3CH_2OH + 2\ CO_2$$

Unfortunately this is currently too energy-intensive to make it a practical method for ethanol production. However, research is underway to make the process better.

In this experiment, you will use a gas sensor to monitor the production of carbon dioxide as the yeast respires with various feedstocks (sugar, ground corn kernels, ground corn stover, and various others). The carbon dioxide produced equals the ethanol produced. You will study and compare the various fermentation rates. You will see first hand the challenge of making ethanol from cellulosic material.

MAKING ETHANOL FROM SUGAR

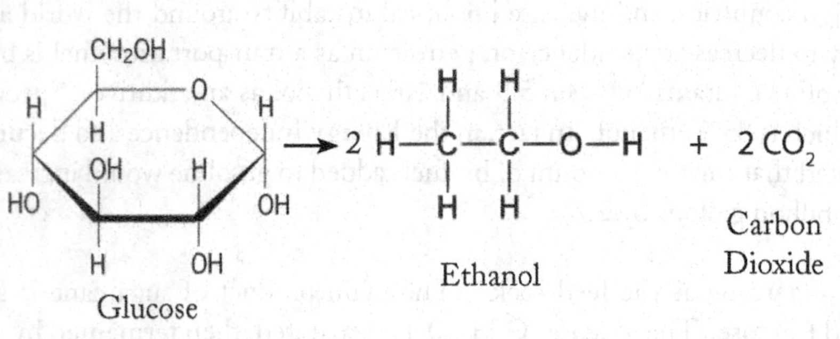

MAKING ETHANOL FROM CORN

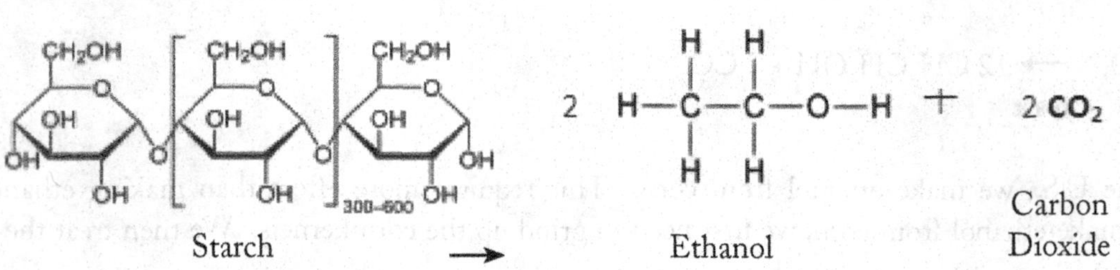

Laboratory 8: Biofuels: Producing Ethanol from Various Feedstocks

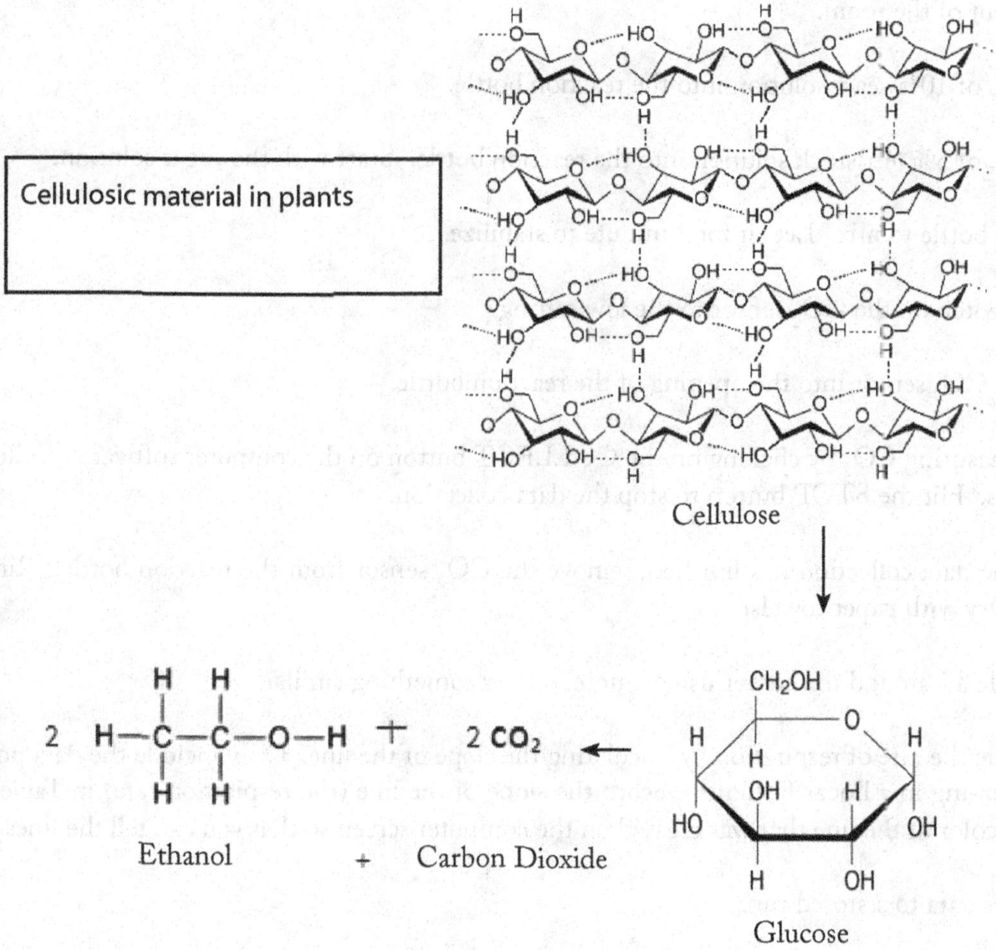

Cellulosic material in plants

Cellulose

2 Ethanol $+$ 2 Carbon Dioxide $\leftarrow$ Glucose

PROCEDURE

Part I: Respiration Rates of Yeast with Various Feedstock

1. Turn on your computer and open the CO_2 sensor software, as directed by your instructor.

2. Plug your CO_2 sensor into port 1 on the top of the handheld device.

3. Plug your handheld device into your computer with a USB cord. The software should automatically recognize the sensor and display CO_2 concentration on the screen.

4. A 10% yeast solution is made by emptying 1 packet of yeast into 100 mL of water in an Erlenmeyer flask. This yeast solution will be made for you and found in the water bath (37-40°C) located in the front of the room.

5. A 5% feedstock solution for each of the feedstock sources is made by putting 5 grams of feedstock into 100 mL of water in a flask. These solutions will be made for you and will be found in the water bath located in the front of the room.

6. Put 2 mL of 10% yeast solution into the reaction bottle.

7. Put 3 mL of 5% feedstock solution into the reaction bottle. Start with the sugar solution.

8. Swirl the bottle to mix. Let sit for 1 minute to stabilize.

9. Set the switch on the CO_2 sensor to the low setting.

10. Place the CO_2 sensor into the opening of the reaction bottle.

11. Begin measuring CO_2 by clicking on the COLLECT button on the computer software. Collect data for 6 minutes. Hit the STOP button to stop the data collection.

12. When the data collection has finished, remove the CO_2 sensor from the reaction bottle. Rinse out the bottle. Dry with paper towels.

13. Fan a little air around the sensor using a notebook or something similar.

14. Determine the rate of respiration by calculating the slope of the line. Only include the data points where it is increasing in a linear fashion. Record the slope of the line (the respiration rate) in Table 8.1, along with the color of the line that was created on the computer screen so that you can tell the lines apart later.

15. Move this data to a stored run.

16. Repeat steps 6-15 for the other feedstock. You MUST test sugar, corn kernels (ground), corn stover (ground), and water. You also need to do one other feedstock of your choice.

17. Write your results on the board so that we can calculate a class average.

18. Record the class average data in Table 8.2.

TABLE 8.1. *Respiration Rates*

FOOD SOURCE	COLOR OF LINE ON GRAPH	RESPIRATION RATE (PPM/SEC)
Sugar		
Corn Kernels (ground)		
Corn Stover (ground)		
Water (control)		

FEEDSTOCK	RESPIRATION RATE (PPM/SEC)
Sugar	
Corn kernels (ground)	
Corn Stover (ground)	
Dark Mulch (ground)	
Light Mulch (ground)	
Raw Sugar	
Corn Starch	
Corn Meal	
Water	

QUESTIONS

1. Which feedstock resulted in the highest respiration rate?

2. Which feedstock resulted in the lowest rate?

3. Does the data follow the trend that you expected?

4. Discuss your results in comparison to what you know about how ethanol is produced to make biofuel.

5. Why are we measuring CO_2 instead of ethanol? What can we infer about ethanol production based on monitoring CO_2 production?

Part II: Breaking Down Corn Stover with Enzymes

1. Make a 5% solution of corn kernels (ground) by measuring 5 grams of corn kernels (ground) and putting it in 100 mL of water.

2. Boil the corn kernel solution on a hotplate for 10 minutes. Cool to 37°C.

3. Put 5 mL of the solution with 2 mL of 1% amylase enzyme solution into the reaction bottle. Put in water bath at 37-40°C for 30 minutes.

4. Run fermentation experiment as in Part I Steps 6-15.

5. Run fermentation experiment as in Part I Steps 6-15 using only water and amylase solution as the feedstock.

6. To break down corn stover with cellulase enzyme, we need to incubate a 5% corn stover solution with 1% cellulase solution and incubate for 3 days. This has already been done for you.

7. Run fermentation experiment on the corn stover with cellulase enzyme as in Part I Steps 6–15.

8. Run fermentation experiment as in Part I Steps 6-15 using only water and cellulase solution as the feedstock.

TABLE 8.3. *Using Enzymes to Break Down Feedstock*

FEEDSTOCK	COLOR OF LINE ON GRAPH	RESPIRATION RATE (PPM/SEC)
Water		
Water and Amylase		
Amylase and Corn Kernels		
Water and Cellulase		
Cellulase and Corn Stover		

QUESTIONS

1. What impact did pretreating the feedstock material with enzymes have on the respiration rate of the yeast with the corn kernels? With the corn stover?

2. What are challenges to making ethanol from corn?

3. What are challenges to making ethanol from corn stover?

LABORATORY 9

Water Quality

OBJECTIVES

1. Learn the nine criteria of water quality and how to measure each
2. Understand the Water Quality Index for interpreting water quality data

INTRODUCTION

Freshwater resources are important to monitor because we use them for so many purposes: drinking, waste removal, transportation, and even recreation. When water becomes "dirty" it can have disastrous health effects; throughout history when water quality was not protected human health has suffered from large scale epidemics.

In the United States, the Water Quality Act (originally passed in 1965) was enacted to give government the authority to regulate how freshwater resources are used, and to enforce guidelines of water quality.

WATER QUALITY INDEX

Developed in 1970, The Water Quality Index (WQI) was a tool for standardizing health evaluations of rivers, streams, and lakes across the country. The WQI consists of nine individual tests: dissolved oxygen (DO), fecal coliform, pH, biochemical oxygen demand (BOD), temperature, total phosphate, nitrates, turbidity, and total solids.

These nine tests are not measured in the same units (e.g. pH is its own scale, DO & BOD are reported in percent saturation; nitrates and phosphates are in mg/L, etc.). The WQI is a system that allows all nine tests to be translated into a single unit called a Q-value. When examining the health of water, some of the nine tests have a stronger impact on overall health than others so the WQI "weights" each of the nine tests based on how strong the impact is (Table 9.1). The final step of the WQI is to sum the final values into one overall number that is evaluated by a group of ranges (Table 9.2).

TABLE 9.1. *An Example of a WQI Table where Raw Values for the Nine Tests have been Converted into a Single Unit (Q-Value), then Multiplied by Their Weighting Factors*

TEST RESULTS	RAW DATA	Q-VALUE	WEIGHTING FACTOR	TOTAL
DO	40% Sat.	30	0.17	5.10
Fecal Coliform	10 colonies/100 mL	80	0.16	12.80
pH	7.5 units	90	0.11	9.90
BOD	10 mg/L	35	0.11	3.85
Temp. Change	5°C	70	0.11	7.70
Total Phosphate	4 mg/L	18	0.10	1.80
Nitrates	0.5 mg/L	75	0.10	7.50
Turbidity	20 NTUs	60	0.08	4.80
Total Solids	250 mg/L	68	0.07	4.76
Overall WQI				**58.21**

TABLE 9.2. *The Categorical Ranges for Overall WQI*

WATER QUALITY INDEX RANGES	
90 – 100	Excellent
70 – 90	Good
50 – 70	Medium
25 – 50	Bad
0 – 25	Very Bad

NINE WATER QUALITY TESTS

Dissolved Oxygen

Oxygen in water is critical to support life; most of the plants and animals living in water need oxygen to "breath" the same way you do! It is the test weighted most heavily in the WQI because it is so imperative to maintaining a healthy body of water. It is mainly supplied by the atmosphere through gas exchange at the surface of water; it also comes from photosynthesizers (e.g. algae and protists).

Dissolved Oxygen (DO) is also influenced by temperature, so taking the temperature of water is an important step in determining the levels of DO. Temperature affects DO because cooler waters can dissolve more oxygen than warm waters. DO measurements are taken with temperature using a chart (provided separately by your instructor) that converts your DO reading from mg/L to percent saturation.

     Laboratory 9: Water Quality

Fecal Coliform

Fecal coliform are a type of bacteria that are not a threat to human health, in fact they are found throughout the intestines of mammals. Because these bacteria feed on mammal waste, their presence in water can be used as an indirect measure of human or other mammal waste. It is this waste that can cause illness in humans or animals if exposed to waters where the waste level is high.

pH

Water contains both H⁺ (hydrogen) and OH⁻ (hydroxyl) ions. The pH test measures the H⁺ ion concentration of aqueous solutions with a scale 0-14; a pH of 7 indicates an equal amount of H⁺ and OH⁻ ions, anything below 7 is acidic, and anything above 7 is alkaline (Figure 9.1).

Cells of most living organisms require a pH of 6-8, so in order to support life, bodies of water must also maintain a pH near this range. Human activity can alter water pH when our waste acts as an input of acids or bases. For example, coal-burning power plants release high amounts of sulfur oxides (SO_x) into the atmosphere, which can be carried to waters through acid rain.

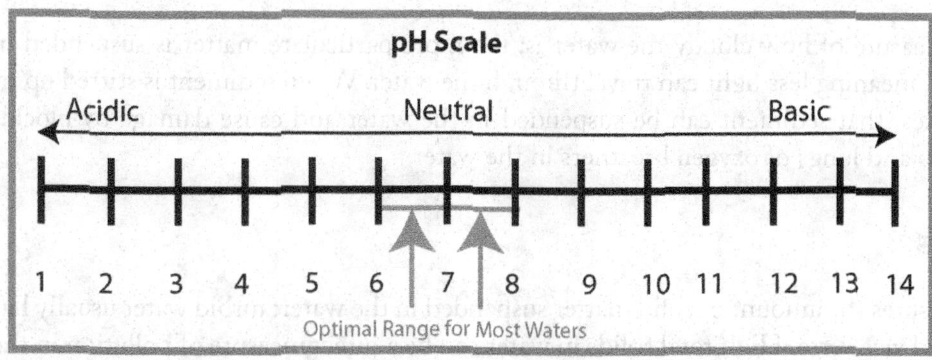

FIGURE 9.1. *pH Scale*

Biochemical Oxygen Demand

Just as DO informs us how much oxygen is present in the water, BOD can inform us how much oxygen is being used in the water. Oxygen is used by all organisms that perform aerobic respiration (oxygen-breathing) but in water quality BOD is specifically used to as a measure of a particular group of oxygen-breathers: decomposers. Decomposers break down organic matter and use oxygen in the process. When BOD is high, it can be an indicator that decomposer populations are high, which is an indication that organic matter levels are also high. Sources of organic matter can be plant material from aquatic plants or from terrestrial plants when plant debris and fertilizers are washed into the water by precipitation (non-point pollution), but can also come from human sources of waste such as industrial by-products or sewage that is dumped directly into bodies of water for disposal (point pollution).

Temperature

Water temperature can determine how much of a substance can be dissolved in the water. Cooler waters can hold more oxygen whereas warmer waters can hold more particulate matter.

Temperature also affects metabolism (think about why we store produce in the refrigerator!); warmer waters can increase rates of photosynthesis and plant growth, which can lead to eutrophication (a process that can deplete water of oxygen). Because of temperature's direct effect on metabolic rates in water, high temperatures are sometimes considered to be a pollutant!

Phosphate and Nitrates

Phosphate and nitrate are forms of macronutrients and can encourage plant growth in water. These two nutrients are also used on land as fertilizers to promote agricultural production and can be washed into water sources. If waters receive too much of these nutrients, water plants (e.g., algae) can accelerate to unhealthy levels causing "scummy" conditions, blocking out sunlight to deeper waters, and eventually leading to eutrophication.

Turbidity

Turbidity is a measure of how cloudy the water is; the more particulate matter is suspended in the water the more turbid it is, meaning less light can travel through the water. When sediment is stirred up from the bottom of rivers and lakes, that sediment can be suspended in the water and cause damage by blocking out light or clogging the gills and lungs of oxygen breathers in the water.

Total Solids

Total solids measures the amount of solid matter suspended in the water: turbid water usually has high amounts of total solids and vice versa. High total solids in water can be a direct measure of pollution in the water because water solids provide a surface that can carry many toxic substances.

1. You will be working in groups of 3.

2. Your group will perform each of the tests 3 times.

3. You will use the water quality test probes to perform the following 6 tests today. Specific directions on how to use the probes will be provided separately.
   - DO (using the CLEAR BOD bottles)
   - pH (using the pH meter)
   - Temperature
   - Turbidity
   - Total Solids
   - Nitrates

4. You will perform the following 2 tests today separately from the test probes:
   - Fecal Coliform (prepare plates today)
   - Phosphorus (prepare and analyze using spectrophotometer)

5. You will perform the following 2 tests next week:
   - BOD (Take reading with the DO water quality test probe next lab in the BLACK BOD bottles)
   - Fecal Coliform (take reading of plates)

6. Record your results on the data table.

7. Calculate the Water Quality Index on the Results Table below using the Q-value charts provided by your instructor.

8. Answer the questions.

DATA SHEETS

Names: ____

Date: ____________________   Time: __________

Testing Location: ___________________________________

Weather Conditions: ______________________________

TEST 1: Dissolved Oxygen (DO)

Temperature Reading (in °C) = ________________________*(this is the only place you need to record temp data)*

DO values: __________   ____________   __________

Take the average of all readings:______________________

Dissolved Oxygen = ____________ mg/L

% Saturation (use the chart & instruction provided by the instructor) = __________________

TEST 2: Fecal Coliform

Sample 1:   __________ colonies/100mL

Sample 2:   __________ colonies/100mL

Sample 3:   __________ colonies/100mL

Fecal coliform (Do not average! Report the highest value!) = __________colonies/100mL

TEST 3: pH

pH values: __________   ____________   __________

Take the average of all the readings:

pH average = ___________________

TEST 4: BOD (will perform next lab!)

Re-record from prior lab: DO = _________mg/L

<u>Dissolved Oxygen Test Results Today:</u>

DO values: _________   __________   __________

Take the average of all the readings:

Today's Dissolved Oxygen = ___________ mg/L

BOD = prior DO: _________mg/L **minus** DO (today): _________mg/L = __________mg/L

TEST 5: Temperature Change

Location 1 (upstream) temperature: _________________ (°C)

Location 2 (downstream) temperature: _________________ (°C)

Temp 1: __________(°C) **minus** Temp 2: ____________(°C) = ___________(°C)

TEST 6: Total Phosphates

___________ mg/L
___________ mg/L
___________ mg/L

Take the average of all readings:

Total Phosphate average = _________________ mg/L

TEST 7: Nitrates

___________ mg/L
___________ mg/L
___________ mg/L

Take the average of all readings:

Nitrate average = _________________ mg/L

<u>**TEST 8: Turbidity**</u>

________________   ________________   ________________

Take the average of all the readings:

Turbidity average = ________________

<u>**TEST 9: Total Solids – using conductivity**</u>

________________   ________________   ________________

Take the average of all the readings:

Conductivity average = _______________________ mg/L

Use the Q-value charts provided by your instructor to translate raw data into Q-values. Then multiply each by weighting factor.

TEST RESULT	RAW DATA	Q-VALUE	WEIGHTING FACTOR	TOTAL
Dissolved Oxygen	% Sat.		0.17	
Fecal Coliform	Colonies/100 mL		0.16	
pH	Units		0.11	
BOD	mg/L		0.11	
Temperature Change	°C		0.11	
Total Phosphate	mg/L		0.10	
Nitrates	mg/L		0.10	
Turbidity	NTUs		0.08	
Total Solids	mg/L		0.07	

Overall WQI (Sum all Total values together)   ________________

OVERALL WQI   **Overall WQI Rating:** ________________
Excellent   90 – 100
Good   70 – 90
Medium   50 – 70
Bad   25 – 50
Very Bad   0 – 25

1. Based on your data, what problems do you see in the Chicago River?

2. Based on your data, how is the Water Quality of the Chicago River?

LABORATORY 10

Alternative Energy: Fuel Cells

OBJECTIVES

1. Understand the process of electrolysis using solar energy
2. Understand how fuel cells work
3. Understand oxidation-reduction reactions

INTRODUCTION

The fuel cell used in this experiment is a Proton Exchange Membrane (PEM) fuel cell. The PEM fuel cell is like a battery in that it creates electricity through a chemical reaction that involves the transfer of electrons. This is an oxidation-reduction reaction. These types of chemical reactions can be split into two parts: the oxidation reaction and the reduction reaction. These are called half-reactions. In the oxidation half-reaction, electrons are released. In the reduction half-reaction, electrons are accepted. A common mnemonic to remember this is OIL RIG (Oxidation Is Losing, Reduction Is Gaining). In the PEM fuel cell, the half-reactions are as follows:

Oxidation: $H_2 \longrightarrow 2H^+ + 2e^-$

Reduction: $4H^+ + O_2 + 4e^- \longrightarrow 2H_2O$

These reactions can be added so that the overall reaction is:

Overall: $2H_2 + O_2 \longrightarrow 2H_2O + energy$

These half-reactions occur at electrodes (a conductor through which electricity passes). In the PEM fuel cell, there are two electrodes: an anode and a cathode. Oxidation occurs at the anode. Reduction occurs at the cathode. A common mnemonic to remember this is that oxidation and anode both begin with a vowel, while reduction and cathode both begin with a consonant. So in the PEM fuel cell, at the anode, hydrogen gas is oxidized and electrons released into the circuit. At the cathode, oxygen gas is reduced and water is formed. In the PEM fuel cell, a proton exchange membrane separates the two electrodes. This membrane allows protons (H^+) to flow through, but prevents electrons from entering the membrane. Thus the electrons are forced to flow through the electrical circuit. See Figure. 10.1.

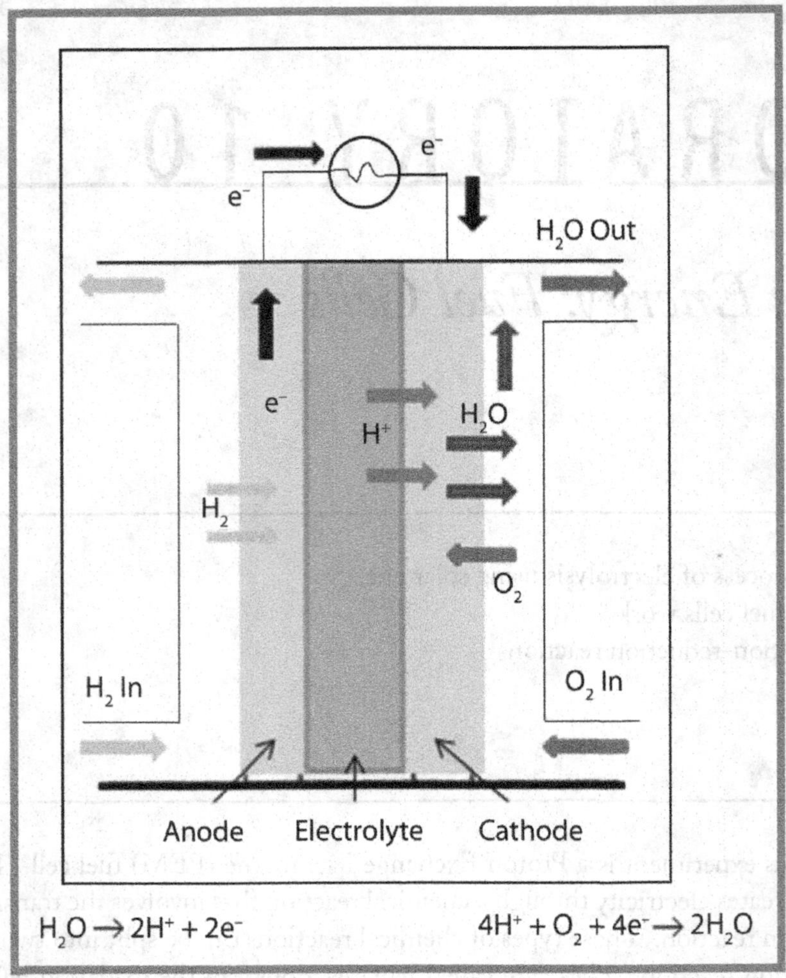

FIGURE 10.1. *PEM Fuel Cell (Modified from U.S. Dept. of Energy)*

We can take advantage of this electrochemical reaction to make cars run on hydrogen gas (H_2) instead of fossil fuels. If we put a PEM fuel cell in a car, we can use the electricity to make the motor run. The only exhaust would be water (H_2O). From an air pollution perspective, this would be advantageous. Unfortunately, H_2 gas does not occur naturally on the Earth. In order to use H_2 as a fuel for cars, we would first need to make H_2. Currently, the most popular way to make H_2 gas is from fossil fuels. This does not help reduce our dependence on fossil fuels, plus it is energy-intensive. Another, less used, method is by electrolysis of water. This also requires an energy source, but it could be a renewable source like wind or solar power. In electrolysis, water (H_2O) is split into its component parts hydrogen gas (H_2) and oxygen gas (O_2) through the following electrochemical reaction:

$$2\,H_2O_{(l)} \longrightarrow 2\,H_{2(g)} + O_{2(g)}$$

<u>Notice that there are twice as many hydrogen molecules produced as oxygen molecules.</u> This reaction does not happen spontaneously, and needs a source of electrical energy, for example, a solar panel. Like the PEM fuel cell reaction, the hydrolysis reaction can be broken up into two half-reactions.

$$\text{Oxidation: } 2H_2O_{(l)} \longrightarrow O_{2(g)} + 4H^+_{(aq)} + 4e^-$$

$$\text{Reduction: } 2H^+_{(aq)} + 2e^- \longrightarrow H_{2(g)}$$

Some automakers are taking advantage of this technology to make cars. One example is the Honda FCX Clarity. The benefits of using fuel cell technology are that it has better efficiency than an internal combustion engine, has zero emissions, and is quiet. The drawbacks are we need to provide more infrastructure, reduce cost, reduce public safety concerns of potential H_2 gas explosions in an accident, and increase use of renewable energy sources when making the H_2 gas.

In this experiment, you will generate hydrogen gas by using a solar panel as the electricity source for hydrolysis of water. Once the H_2 gas is produced, you will use the fuel cell to make your car go!

PROCEDURE

Part I: Electrolysis – Creating Hydrogen from Water Using the Solar Panel

1. Set up the fuel cell car as described on the package instruction sheet.

2. Once the car is set up, insert one end of the red cable into the solar panel and the other end into the fuel cell. Be sure to put it in the red jack on the solar panel and the oxygen side (red side) of the fuel cell.

3. Insert one end of the black cable into the solar panel and the other end into the fuel cell. Be sure to put it in the black jack on the solar panel and the hydrogen side (black side) of the fuel cell.

4. Shine the light directly on the solar panel. You will soon see oxygen and hydrogen gas bubbling out into the gas cylinders on the back of the car. It will take about 5-10 minutes to completely fill the hydrogen cylinder. Watch the amount of each gas that is produced and use this information to fill in the Tables 10.1 and 10.2.

5. When the hydrogen cylinder is full, disconnect the solar panel from the fuel cell. You can tell that the hydrogen cylinder is full when all of the water is forced out of the hydrogen cylinder and you can see it is only filled with gas.

Part II: Using Hydrogen to Make the Car Run

1. Make sure the fuel cell is disconnected from the solar panel.

2. Plug the red and black wires from the motor into the red and black jacks on the fuel cell.

3. Place the kit on a smooth and flat surface and watch it go! The two blue LED lights on the front of the motor will also start to flash.

4. The car is like a Roomba vacuum and will independently find its way past any obstacles in its path. If it hits a barrier, it will turn and reverse until it can find a forward direction clear of barriers.

5. The car will continue to run on its own until all the hydrogen gas stored in the inner cylinder is consumed. RECORD THE LENGTH OF TIME YOUR CAR RUNS ON THE HYDROGEN GAS.

DATA: ELECTROLYSIS

TABLE 10.1. *HYDROGEN*

TEST TIME (SEC)	GAS VOLUME GENERATED (ML)	TEST TIME (SEC)	GAS VOLUME GENERATED (ML)
0		210	
30		240	
60		270	
90		300	
120		330	
150		360	
180		390	

TABLE 10.2. *OXYGEN*

TEST TIME (SEC)	GAS VOLUME GENERATED (ML)	TEST TIME (SEC)	GAS VOLUME GENERATED (ML)
0		210	
30		240	
60		270	
90		300	
120		330	
150		360	
180		390	

QUESTIONS

1. What is the function of the solar panel?

2. What is the overall reaction for electrolysis of water?

3. Was more hydrogen or more oxygen produced?

4. What was the ratio of hydrogen to oxygen produced? [Take the hydrogen volume generated (ml) and divide by oxygen volume generated (ml).]

5. Theoretically, what should the ratio of hydrogen to oxygen be, based on the chemical reaction?

QUESTIONS: FUEL CELL CAR

1. Was your car able to run on the fuel cell?

2. How long was it able to run on the fuel cell?

3. What is the overall reaction in the fuel cell?

LABORATORY 11

Atmospheric CO$_2$ Concentration

OBJECTIVES

1. Understand trend in atmospheric carbon dioxide concentration over past 50 years
2. Understand trend in atmospheric carbon dioxide over one-year time span
3. Be able to model data and use model to make predictions

INTRODUCTION

According to the Fourth Assessment Report of the Intergovernmental Panel on Climate Change (IPCC), warming of the global climate is unequivocal. This is evident from the observations of increases in global average air and ocean temperatures. Over the past 100 years, the global average temperature has increased 0.74°C. In addition, the linear warming trend over the past 50 years is nearly twice that of the 100-year trend. Changes in atmospheric concentrations of greenhouse gases, especially carbon dioxide, are one of the drivers of this climate change. The annual emission of carbon dioxide has grown by about 80% since 1970. As a result, the global atmospheric carbon dioxide concentration today far exceeds pre-industrial values as seen in ice cores. In fact, the atmospheric concentration of CO$_2$ in 2005 exceeded by far the natural range over the past 650,000 years. In this lab, you will analyze the atmospheric CO$_2$ concentration measured at the Mauna Loa Observatory over the range of 1958 to 2011. This data was collected by the Earth System Research Laboratory (ESRL) of the National Oceanic and Atmospheric Administration (NOAA).

PROCEDURE

1. Open the Atmospheric CO$_2$ Data Excel spreadsheet. (The raw data came from: ftp://ftp.cmdl.noaa.gov/ccg/co2/trends/co2_mm_mlo.txt)

2. Make a scatter plot with smooth line graph of the data. Years should be on the x-axis and atmospheric CO_2 concentration (in ppm) on the y-axis.

3. Format the graph to make a nice presentation of the data.
   a. Set your minimum value for the y-axis to be 300.0 ppm.
   b. Make the graph on its own page (and not as an object within the spreadsheet).
   c. Delete the legend. It is not needed since there is only one data series.
   d. Give your graph a title and label the 2 axes.

4. Make a best-fit line (a trendline) through all 53 years worth of data.
   a. Include the equation of the line on the graph.
   b. Include the R^2 value on the graph.

5. Make a best-fit line (a trendline) including ONLY THE LAST 25 years worth of data (1986-2011).
   a. Include the equation of the line on the graph.
   b. Include the R^2 value on the graph.

6. Make a best-fit line (a trendline) including ONLY THE LAST 10 years worth of data (2001-2011).
   a. Include the equation of the line on the graph.
   b. Include the R^2 value on the graph.

7. Attach a copy of your graph with your report.

QUESTIONS

Use the graph you just created and the data table to answer the following questions.

1. What are the units of the CO_2 concentration? ___________

2. What does 2011.625 mean in terms of month and year? (Note: the decimals are fractions of a year. So .33 = 1/3 = .33*12 months = 4 = April; .75 = ¾ = .75*12 months = 9 = September) ___________

3. Describe the overall shape of the curve.

4. What was the CO_2 concentration in 1958.208? ___________________
   What was the CO_2 concentration in 2011.625? ___________________

   What is the percentage change from 1958.208 to 2011.625? ___________

Laboratory 11: Atmospheric CO_2 Concentration

5. Use your cursor and move it to one of the peaks of the oscillating data. Note the time of year that the maximum occurs. (For example, 1972.25 is ¼ of the way through 1972, so March 1972.)

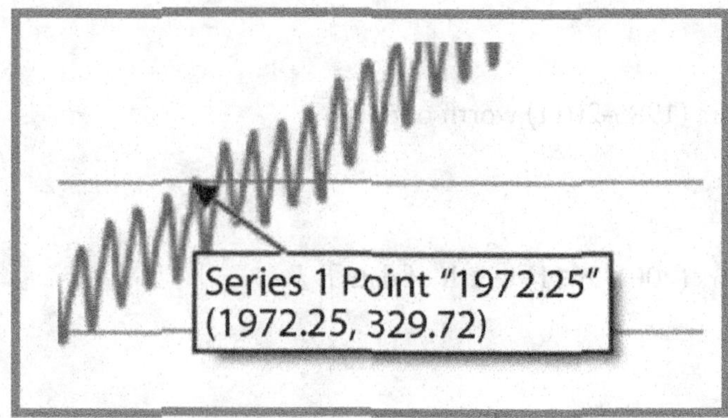

Do the same thing to find out what month the minimum usually occurs.

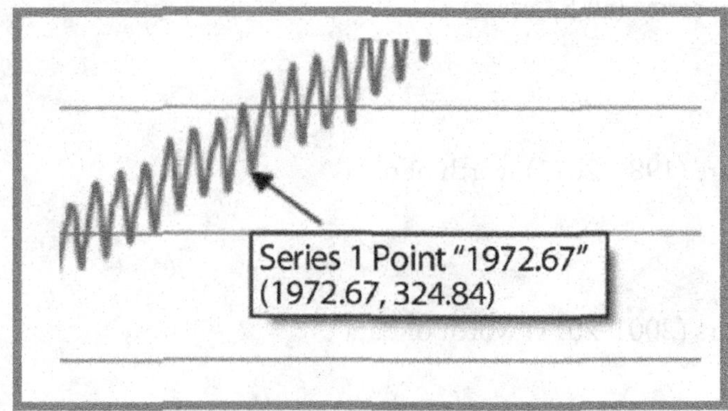

In general, in what months do the maxima and minima CO_2 concentration occur?

Maxima ________________   Minima ________________

6. What do you think causes the atmospheric CO_2 concentration to oscillate during the year?

7. Determine the long-term rate of change of the Mauna Loa CO_2 concentrations. The rate of change is the slope of the straight line that appeared in the graph. The displayed equation gives you this slope: $y = mx + b$ where $y = slope*x + intercept$.

a. What is the equation of your line for ...

   i. The whole data series (all 53 years)

   ii. The last 25 years (1986-2011) worth of data

   iii. The last 10 years (2001-2011) worth of data

b. What is the rate of change (including units) of the Mauna Loa CO_2 concentration (the slope of the line) for ...

   i. The whole data series (all 53 years)

   ii. The last 25 years (1986-2011) worth of data

   iii. The last 10 years (2001-2011) worth of data

c. What does the slope of the line actually mean?

d. How does the slope of the line change when you are looking at the 53-year time span vs. the 25-year time span vs. the 10-year time span?

e. What implications does this have with respect to climate change?

f. <u>Using the equation of the line for the whole 53 years worth of data</u>, what will the CO_2 concentration be in the year 2025? SHOW YOUR WORK!

g. Do you think the ACTUAL value of the CO_2 concentration in 2025 will be greater than or less than what you are predicting? Why?

1. To make a scatter plot of your data:
   a. Highlight all of the data that you want to include on your graph.
   b. Click on "Insert," then "Scatter," then choose the scatter plot with smooth line.

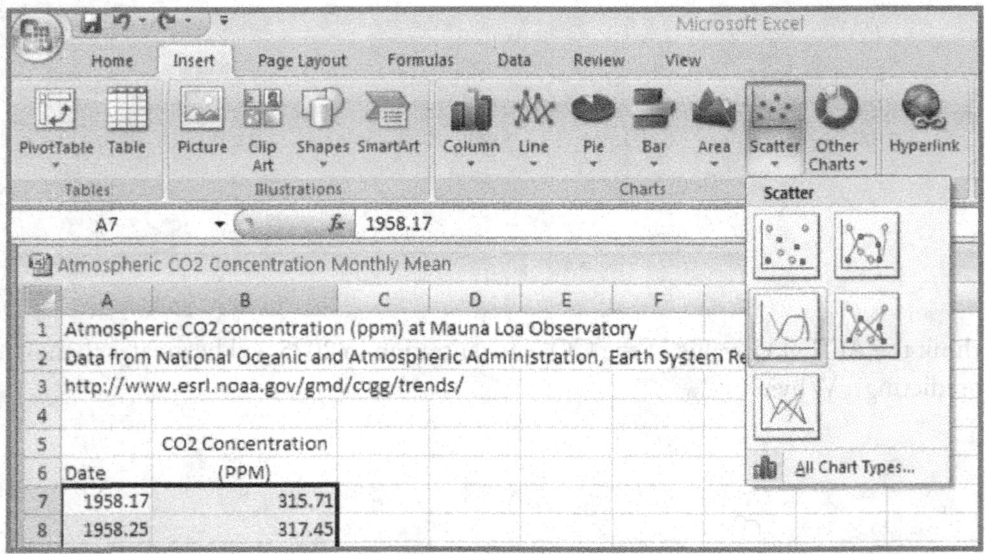

2. To change the minimum value for the y-axis of your graph to 300.0:
   a. Go to "Chart Tools," then "Layout," then "Axes," then "Primary Vertical Axis," then "More Primary Vertical Axis Options";
   b. In the pop-up box, under "Axis Options," set the Minimum at a Fixed 300.0, then "Close."

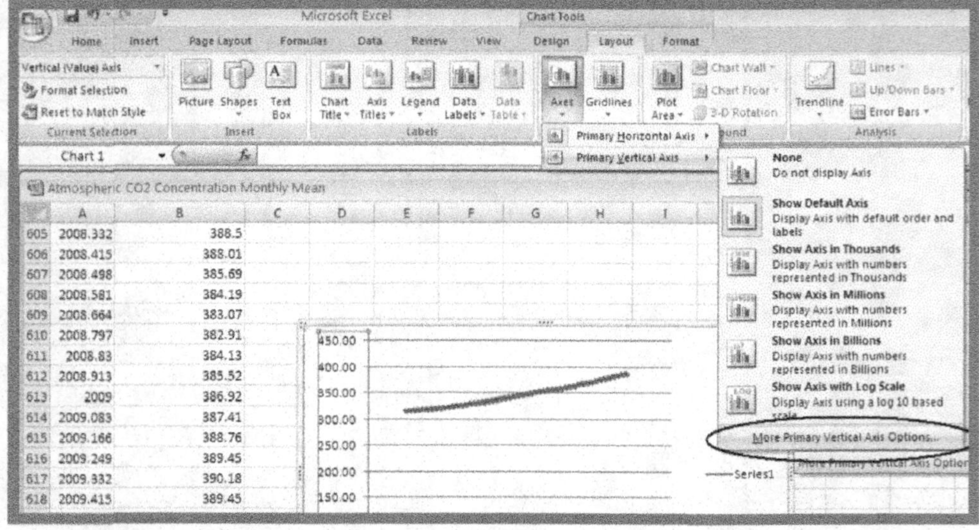

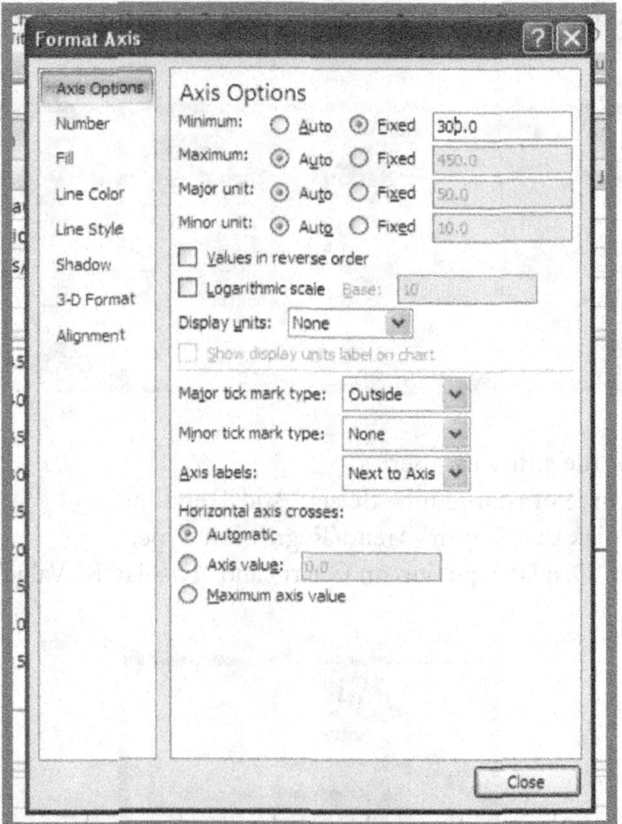

3. To make the graph on its own sheet instead of as an object on the spreadsheet:
   a. Click on "Chart Tools," then "Design," then "Move Chart Location";
   b. In the pop-up box, choose "New Sheet" option;
   c. Click OK.

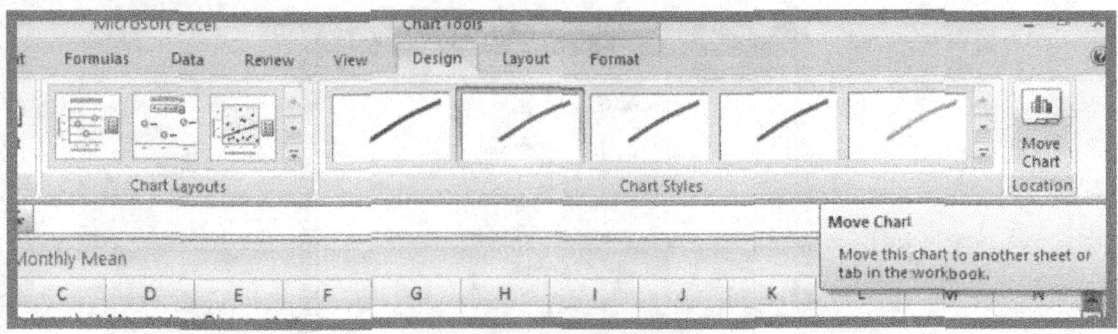

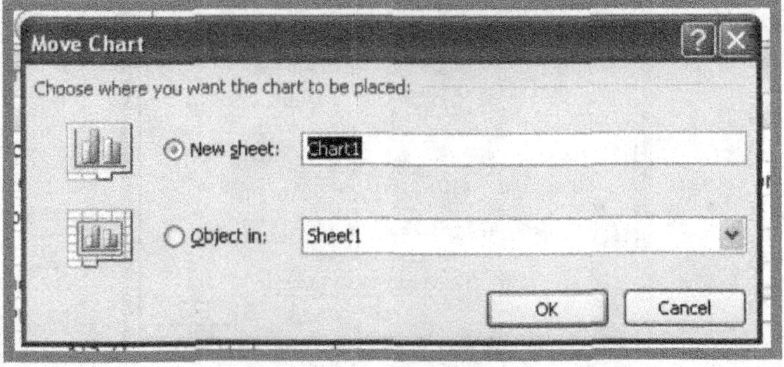

4. To give your graph a nice title and to label the two axes, you should use the "Chart Tools" and "Layout" menus.

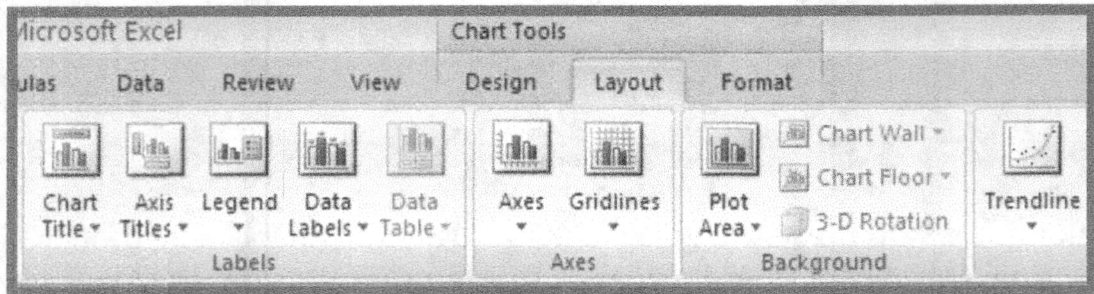

5. To insert a trendline for the entire data series:
    a. Right click on the curve of your graph. Select "Add Trendline";
    b. In the pop-up box, pick the "Linear" Trend/Regression Type;
    c. Make sure you click "Display equation on Chart" and "Display R^2 Value";
    d. Click "Close."

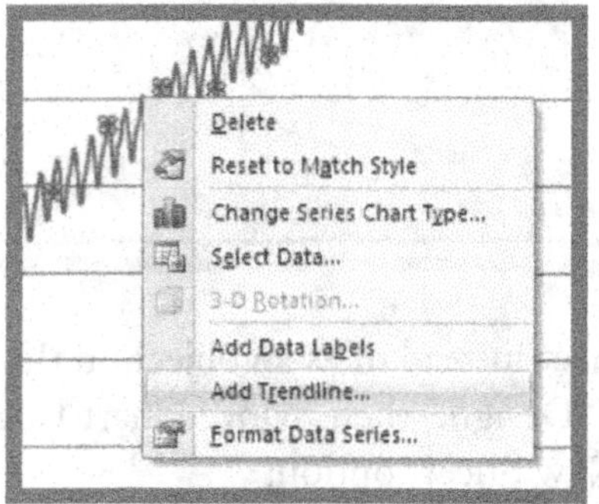

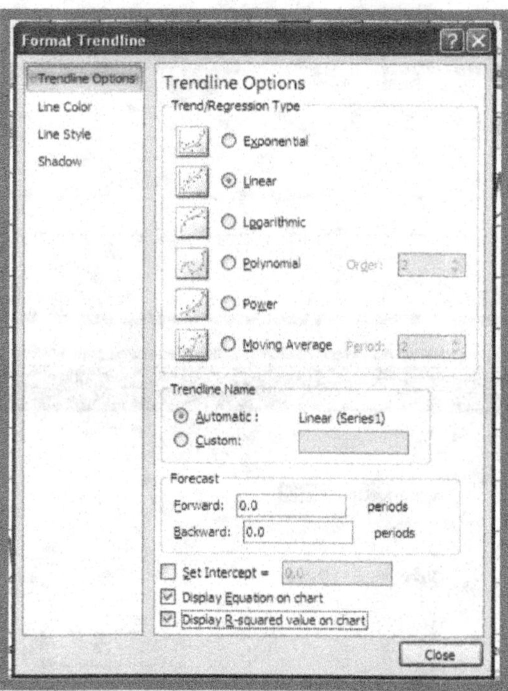

Laboratory 11: Atmospheric CO_2 Concentration

6. To create a localized trendline:
   a. Create an x-y scatter plot of all the data;
   b. Go to "Chart Tools," then "Design" tab, then "Select Data."

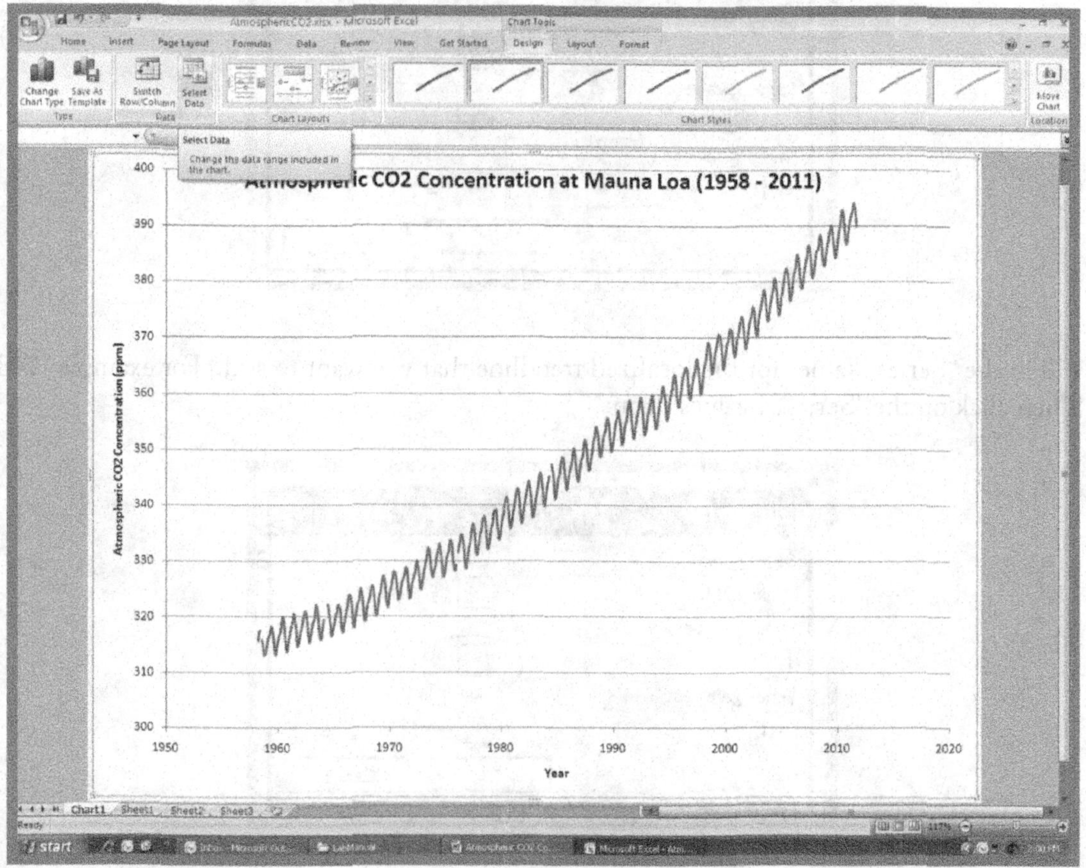

   c. A popup box will appear. Click on "Add" to add a new series.

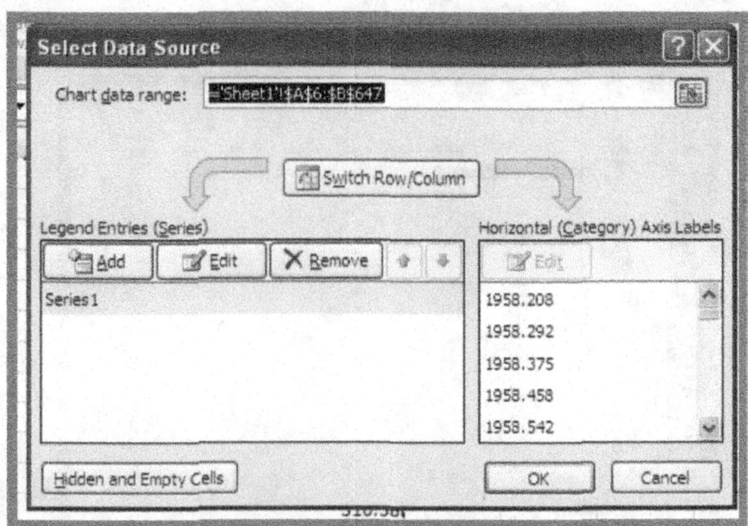

d. A new popup box will appear.

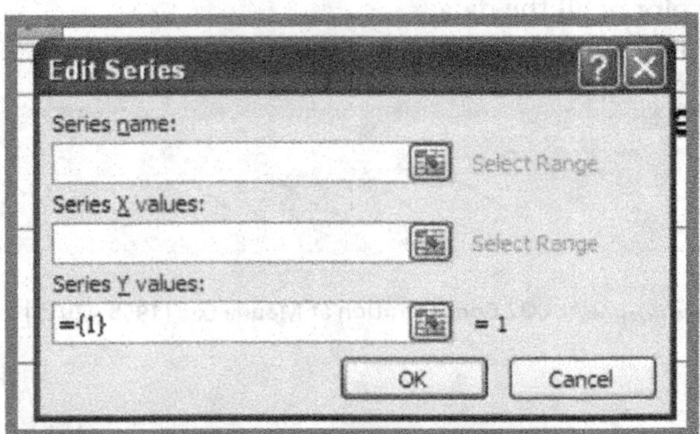

e. Fill in the "Series Name" for the localized trendline that you want to add. For example "1986-2011." Then click on the "Series X values" icon:

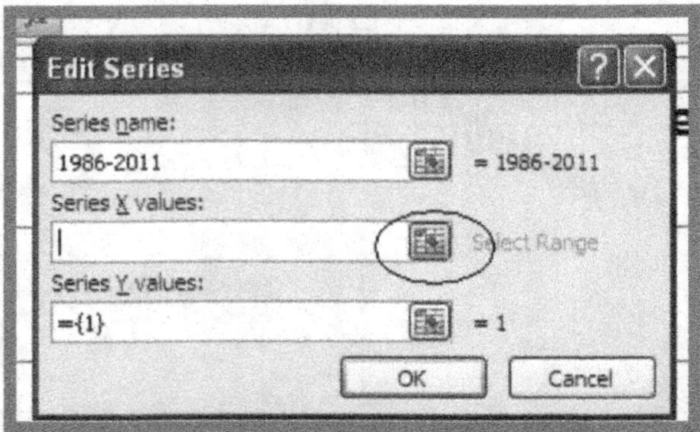

f. A new popup box will appear. Do NOT click on that, but go to the "Sheet 1" tab at the bottom left of your graph. Click on "Sheet 1" and it should take you to the datasheet. Once here, highlight the x-values for your points that you want to graph. For example, if you want to graph all the x values from 1986-2011, highlight all the cells from 1986-2011. Hit "Enter" key.

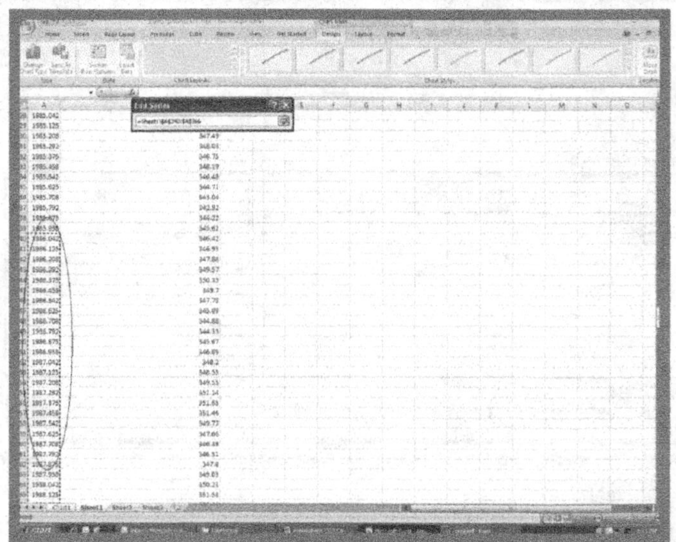

g. The "Edit Series" box should appear again with the X-values listed.

h. Now click on the Series Y values icon.

i. A new popup box will appear. Do NOT click on that, but go to the "Sheet 1" tab at the bottom left of your graph. Click on "Sheet 1" and it should take you to the datasheet. Once here, highlight the y-values for your points that you want to graph. For example, if you want to graph all the y-values from 1986-2011, highlight all the y-axis value cells that correspond to the time frame 1986-2011. Hit "Enter" key.

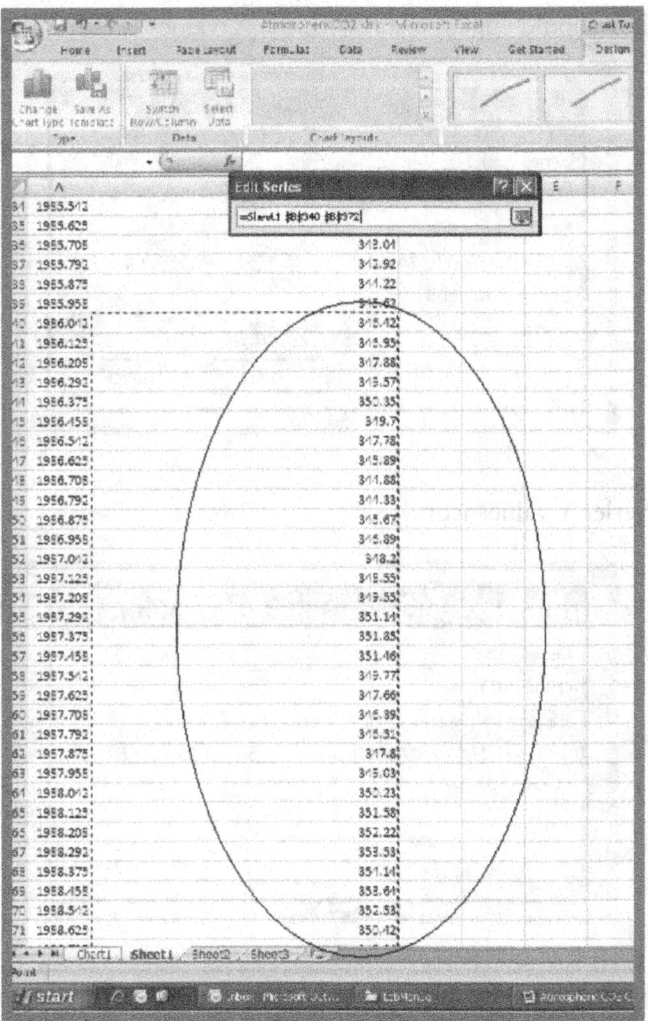

j. You should end up with a box that looks like this:

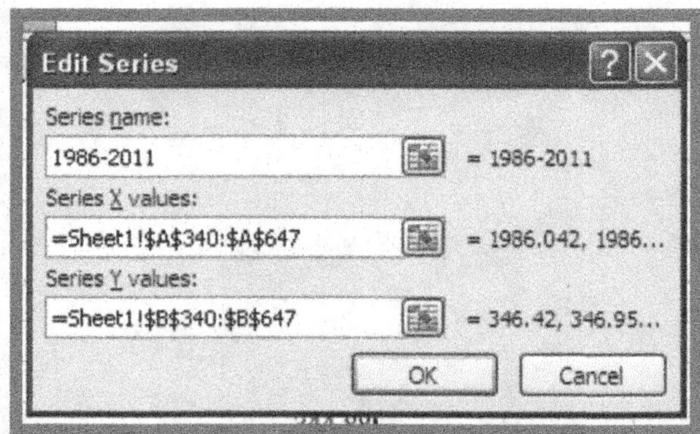

k. Once you click OK, your new series should appear on the list.

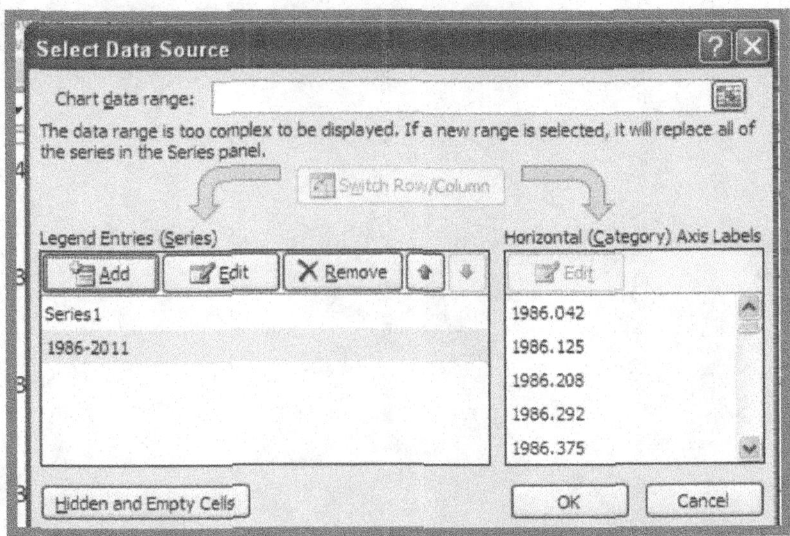

l. Click OK and your new series will appear on your graph (here in the darker shade).

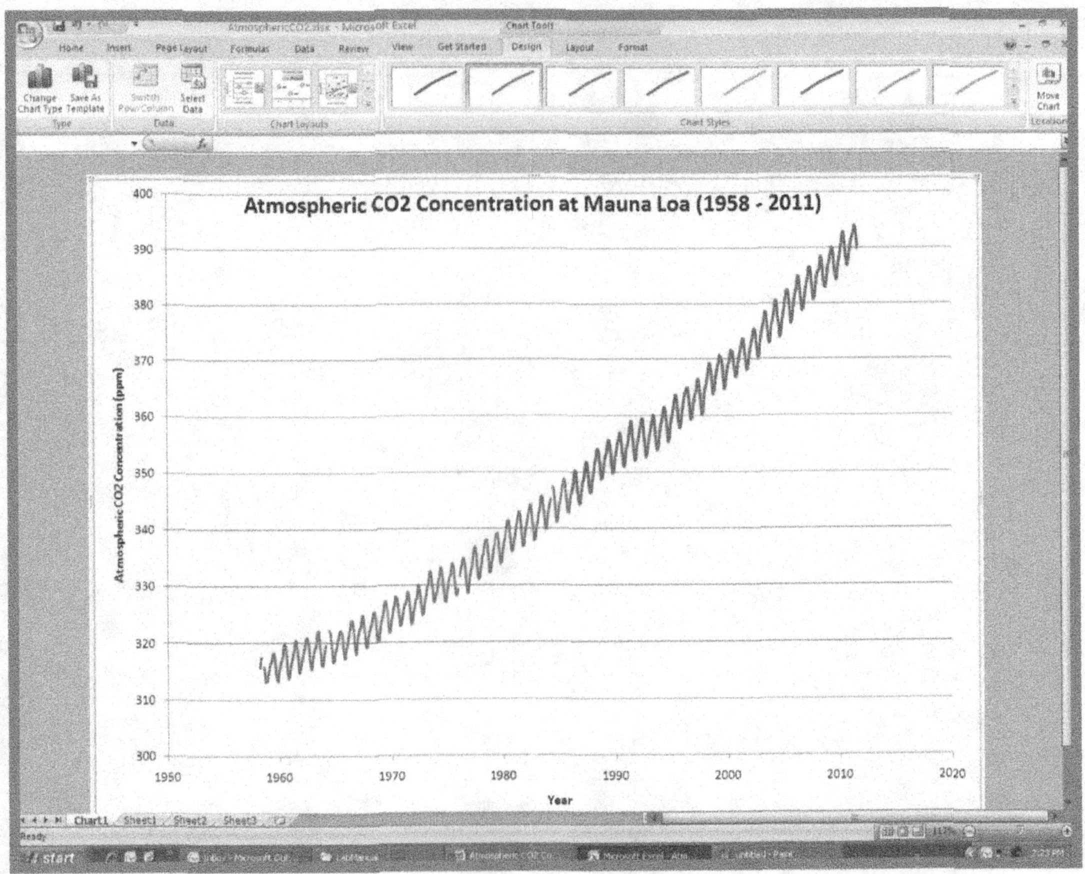

m. Right click on your new series and select **Add Trendline**. Follow the usual steps for inserting a trendline, and a localized trendline will be added to your chart!

LABORATORY 12

The Rock Cycle

OBJECTIVES

1. Identify and classify rocks from samples
2. Describe the rock cycle
3. Compare and contrast metamorphic, igneous, and sedimentary rocks

INTRODUCTION

Rocks are all around us. They are what make up the earth's crust. Rocks are defined as a solid mixture of one or more minerals. Rocks can be classified into three categories: igneous rocks, metamorphic rocks, and sedimentary rocks.

Igneous rocks are formed when molten magma cools and crystallizes. When magma cools slowly deep within the earth's crust, the igneous rocks that form are called **intrusive** and are characterized by a coarse-grained texture. Coarse-grained rocks are those in which you can see the individual mineral grains without using a magnifier. The second major group of igneous rocks are called **extrusive** rocks. These rocks form when the magma rises and flows out at the earth's surface as lava. This lava will be more quickly cooled to form rocks that are much finer grained. Due to the fine grains, you need a magnifying class in order to see the individual mineral grains.

Sedimentary rocks can be classified into two main categories based on how they are formed: physical or chemical sedimentation. **Clastic** sedimentary rocks are composed of sediments that were transported by moving fluids (water, wind) and physically deposited when these fluids came to rest. These sediments can include clay, silt, sand, and gravel. **Non-clastic** sedimentary rocks form rock layers when dissolved materials precipitate from solution or with lithification of once-living matter. Limestone is a prime example of chemical sedimentary rocks.

<u>**Metamorphic rocks**</u> are created when pre-existing rocks are modified by heat, pressure, and chemical processes while deep in the Earth's crust. These rocks widely vary, depending on the composition of their parent rock and the degree of metamorphism. During metamorphism, the minerals in the parent rock may be bent and crushed, or they may be completely altered to new materials. Metamorphic rocks are classified by whether or not they are foliated (layered or banded). Those rocks that have distinct bands or layers are called **foliated** metamorphic rocks. Those rocks that do not have distinct bands or layers are called **non-foliated** metamorphic rocks.

Rocks are continually changing. Water, wind, and weather are continually breaking them down. This breaks little pieces of sediments off of the rocks. As the sediment is buried, it is compressed together to make it into sedimentary rock. When a great amount of pressure along with high temperatures is exerted onto sedimentary rock, it can turn into a metamorphic rock. If these metamorphic rocks are deep in the earth's crust and are exposed to high enough temperatures, they melt. If this molten magma reaches the earth's surface and cools, it will crystallize to form igneous rocks. This is an example of the **Rock Cycle**.

PROCEDURE

Part I. Rock Cycle Video

1. Watch the video on the rock cycle.
2. Record answers to questions on the Data Sheet: Part I as the video is playing.
3. Discuss as a class the questions from the video.

Part II. Examination of Rocks

Examine the rock collection and fill in Data Sheet: Part II with a description (color, texture, weight/density, distinguishing features) of the rock and indication of how it was formed. Pay close attention to SIMILARITIES within each rock type and DIFFERENCES between each rock type.

1. Igneous
2. Sedimentary
3. Metamorphic

Part III. Rock Identification

1. Work in groups of 3.
2. Follow the ID key to identify each of the 15 rocks listed.
3. Fill in the answers on the Data Sheet: Part III.

DATA SHEET: PART I. ROCK CYCLE VIDEO

1. What are the 3 basic families of rocks?

2. How is granite formed?

3. How is basalt formed?

4. What is common between granite and basalt?

5. What happens to rocks as they travel downstream?

6. Name different ways sediment can be deposited.

7. What happens to the lowest layer in a sediment basin?

8. Describe what a conglomerate rock looks like.

9. If a rock has crystals, what does it mean?

10. Name the daughter rock of …

    Granite →

    Limestone →

    Conglomerate →

11. What makes a rock metamorphic?

12. What powers the rock cycle?

DATA SHEET: PART II. EXAMINATION OF ROCKS

IGNEOUS ROCKS

NAME	DESCRIPTION	HOW WAS IT FORMED?
Obsidian		
Basalt		
Pumice		
Biotite Granite		
Gabbro		

SEDIMENTARY ROCKS

NAME	DESCRIPTION	HOW WAS IT FORMED?
Coquina		
Quartz Conglomerate		
Red Sandstone		
Bituminous Shale		
Shell Limestone		

METAMORPHIC ROCKS

NAME	DESCRIPTION	HOW WAS IT FORMED?
Granitoid Gneiss		
Schist		
Gray Slate		
White Marble		
Soapstone		

DATA SHEET: PART III. ROCK IDENTIFICATION

Use the ID Key on the following pages to identify each rock listed below. **Write the number that is on the rock in the appropriate space once you have identified each rock.**

<u>Igneous Rocks</u>   <u>Sedimentary Rocks</u>   <u>Metamorphic Rocks</u>

Granite _____   Limestone _____   Gneiss _____

Basalt _____   Coquina _____   Marble _____

Gabbro _____   Shale _____   Schist _____

Obsidian _____   Conglomerate _____   Soapstone _____

Pumice _____   Sandstone _____   Slate _____

Igneous Rocks (orange set)

Step 1 – Look at the rocks in the igneous rock set. Separate the coarse-grained rocks form the fine grained rocks. You should have 2 coarse-grained rocks and 3 fine-grained rocks.

Step 2 – Look at the set of 2 coarse-grained rocks. These are **intrusive** igneous rocks which means that they cooled slowly deep within the earth's crust. Find the rock that is light in color. It will have white, gray, and pink minerals. These are quartz, feldspar, and biotite. This rock is **granite**. The remaining sample will be dark in color with no light minerals. It is dark gray to black. This rock is **gabbro**. It is made up of feldspar and clinopyroxene. This is the most abundant rock in deep, oceanic crust.

Step 3 – Look at the set of 3 fine-grained rocks. These rocks were formed when lava was cooled quickly on the earth's surface, thus the fine grains. These are the **extrusive** igneous rocks. The rock that is black with a glassy appearance is **obsidian**. It is often called volcanic glass.

Now look for the rock that is light in color with cells/holes. This rock is very lightweight. It is called **pumice**. Pumice forms from lava that has a high content of volcanic gases is thrown out of a volcano. As the lava cools quickly, the gases escape and the rock is left with a frothy texture.

The last sample is a fine-grained, gray rock. It is called **basalt**. It is similar in composition to gabbro; however, the rock cooled quickly, therefore the fine grains. Basalt underlies more of the earth's surface than any other rock.

Sedimentary Rocks (black set)

Step 1 – Look at the set of sedimentary rocks. We will use an acid test to separate these rocks into ones formed by physical sedimentation (clastic rocks) and ones formed by chemical sedimentation (non-clastic rocks). If the rock reacts with acid, it was formed by chemical sedimentation. If it does not react with acid, it was formed by physical sedimentation. <u>You must wear goggles and gloves when doing this part of the experiment.</u> Put one or two drops of acid on the rocks. Separate those rocks that reacted with acid from those that did not. You should have 2 that reacted and 3 that did not.

Step 2 – Take those rocks that reacted with acid. These are the **non-clastic rocks**. Look at the rocks. Do you see one that is porous, light-weight and composed of broken shell fragments? This rock is **coquina**. Do you see one that is fine-grained, smooth, and dense? This rock is **limestone**.

Step 3 – Go back to the rocks that did NOT react with acid. There should be 3 of them. These are the **clastic rocks**. Are there grains visible in the rocks? Separate them into those that have grains that are visible and those that have no grains. There should be 2 with grains visible and 1 with no grains visible. The rock with no grains visible is **shale**. It should feel smooth; it will not scratch your fingernail; it might be layered.

Step 4 – Take the 2 rocks that had visible grains. The rock that has rounded, gravel-sized grains is **conglomerate**. The one that has small grains throughout is **sandstone**.

Metamorphic Rocks (purple set)

Step 1 – Look at the rocks in this set. Sort them into 2 groups: those that are foliated (layered) and those that are not. There will be 3 foliated rocks and 2 non-foliated.

Step 2 – Take the set that was layered. These are the **foliated** metamorphic rocks. Do you see one that is distinctly banded (foliated)? It is normally characterized by alternating white and dark bands. This rock is **gneiss**. Next look for a fairly coarse-grained rock. Mica will be clearly recognizable with its sparkly surface. This rock is called a **schist**. The last rock is clearly foliated with a fine-grained, slatey cleavage. It has distinct, smooth, flat sheets. This rock is **slate**. This rock was formed by metamorphic processes from the parent rock shale.

Step 3 – Now look at the subset of 2 rocks that were **non-foliated**. <u>You must wear goggles and gloves when doing this part of the experiment.</u> Put one drop of dilute acid on each rock. One of the samples will react with the acid and will bubble when tested. This rock is **marble**. Metamorphic processes changed a type of limestone into this marble. (Recall that the limestone from the sedimentary set of rocks above also reacted with the dilute acid.)

The remaining sample did not react with the dilute acid. This rock has a slippery feel and is not very hard. You should be able to scratch it with your fingernail and leave a mark. The main mineral of this rock is talc. This rock is called **soapstone**.

1. Draw a picture of the rock cycle.

2. Compare and contrast intrusive and extrusive igneous rocks.

3. Compare and contrast clastic and non-clastic sedimentary rocks.

4. Compare and contrast foliated and non-foliated metamorphic rocks.

5. Give an example of a parent and daughter rock set.

6. Give an example of sediment along with its corresponding sedimentary rock.